THE VARIATIONAL APPROACH TO FRACTURE

THE VARIATIONAL APPROACH TO FRACTURE

Blaise BOURDIN
Louisiana State University

Gilles A. FRANCFORT
Université Paris-Nord

Jean-Jacques MARIGO
Université Pierre et Marie Curie

Blaise Bourdin
Louisiana State University

Gilles A. Francfort
Université Paris-Nord

Jean-Jacques Marigo
Université Pierre et Marie Curie

Library of Congress Control Number: 2008921762

ISBN 978-1-4020-6394-7   e-ISBN 978-1-4020-6395-4

Printed on acid-free paper.

9 8 7 6 5 4 3 2 1

springer.com

Foreword

One of the goals of the *Journal of Elasticity: The Physical and Mathematical Science of Solids* is to identify and to bring to the attention of the research community in the physical and mathematical sciences extensive expositions which contain creative ideas, new approaches and current developments in modelling the behaviour of materials. Fracture has enjoyed a long and fruitful evolution in engineering, but only in recent years has this area been considered seriously by the mathematical science community. In particular, while the age-old Griffith criterion is inherently energy based, treating fracture strictly from the point of view of variational calculus using ideas of minimization and accounting for the singular nature of the fracture fields and the various ways that fracture can initiate, is relatively new and fresh. The variational theory of fracture is now in its formative stages of development and is far from complete, but several fundamental and important advances have been made. The energy-based approach described herein establishes a consistent groundwork setting in both theory and computation. While it is physically based, the development is mathematical in nature and it carefully exposes the special considerations that logically arise regarding the very definition of a crack and the assignment of energy to its existence. The fundamental idea of brittle fracture due to Griffith plays a major role in this development, as does the additional dissipative feature of cohesiveness at crack surfaces, as introduced by Barenblatt.

The following invited, expository article by B. Bourdin, G. Francfort and J.-J. Marigo represents a masterful and extensive glimpse into the fundamental variational structure of fracture. It contains examples from both theory and computation, and it suggests a related, introductory extension into the field of fatigue.

Minneapolis, 2007   Roger Fosdick

Preface

In this tract, we wish to offer a panorama of the variational approach to brittle fracture that has developed in the past eight years or so. The key concept dates back to Griffith and consists in viewing crack growth as the result of a competition between bulk and surface energy. We revisit Griffith's insight in the light of the contemporary tools of the Calculus of Variations. We also import Barenblatt's contributions and always strive to compare the respective merits of both types of surface energy. The advocated variational approach provides a picture of initiation and propagation which we hope to be both thorough and incisive.

The reader will gauge the novelty and appropriateness of the approach by what it delivers, or fails to deliver, on specific issues like those of crack initiation, or crack path. Success or failure are thus very much anchored in the concrete performance of the method. The performance is also modulated by the choice of surface energy, the impact of which we make a concerted effort to distinguish from that of the variational.

The material is mathematical in nature, and the reader should expect theorems and sometimes proofs. However, we are not overly preoccupied with mathematical technicalities, because we do not view the material as a contribution to the field of mathematics, per se. The presence of mathematics illuminates the model and highlights the inconsistencies or difficulties that may arise. Mechanical modeling is an old field and it has matured through its intimacy with mathematical techniques and analysis. We honor that strong tradition, which strongly flavors any treatise on mechanics.

Throughout the text, we have tried our best to connect the approach with more classical treatments of fracture, and to illustrate the results in simple test settings, or through relevant numerical simulations. The proposed approach certainly is rooted in the familiar of fracture, and the model does assume minimal comfort with the rationalization of the thermo-mechanics of solid continua. It does so for clarity of exposition and not out of a strong belief in any mechanical dogma. In this respect, we do implicitly use the framework of what are sometimes called standard generalized materials, because of its compliance with the tenet of rational mechanics. Unfortunately, that seasoned framework becomes hesitant when confronted with non-convexity, an unalterable feature of the energetic landscape of fracture.

Our departure from the classical setting is then to be viewed as a partial completion of that setting. The import of the variational notion of meta-stability has the arguable merit to resolve many of the usual indeterminacies of the resulting models. The mechanically minded reader will hopefully concede that the proposed approach transcends mere fracture and suggests a more general treatment of rate independence. A variational approach to many rate independent processes has been a topic of active research in recent years and we will point to the relevant literature throughout the text.

The introduction details the table of contents. The model is presented in the first three sections (Sections 1 to 3). Its impact on the main issues of fracture is analyzed in the next four sections (Sections 4 to 7). Numerical implementation is the object of Section 8, while the extension of the model to the realm of fatigue occupies Section 9.

As also emphasized in the introduction, the presented work goes far beyond our own contributions and many names are attached to the development of what we think of as "the variational view". We acknowledge those with great pleasure at the close of the introduction.

Support for the first author's work was provided in parts by the National Science Foundation grant DMS-0605320. The second author wishes to acknowledge the hospitality of the Centro di Matematica Ennio De Giorgi, Pisa, during the trimester "Calculus of Variations and Partial Differential Equations", 01 September 2006 - 15 December 2006, during which a large part of this manuscript was written. The numerical experiments were performed using the National Science Foundation TeraGrid resources provided by NCSA under the Medium Resource Allocation TG-DMS060011N.

This tract was born, grew and was completed with Pr. Roger L. Fosdick's strong support and gentle prodding. Without his help and that of the Journal of Elasticity: The Physical and Mathematical Science of Solids, this manuscript would undoubtedly be stillborn. Of course, the obstetrician cannot be held accountable for the mental state of his patient, and the authors assume full responsibility for the errancies that mar the presented material.

Paris, March 2007   B. Bourdin, G. Francfort & J.-J. Marigo

Table of Contents

1. Introduction

At the risk of being unfair, we credit first and foremost A. A. Griffith for developing the field of brittle fracture. His views were that cracks are the macroscopic manifestation of putative debonding at the crystalline level, that this process can be accurately portrayed through an energy density at each point of the crack surface and that crack propagation results from the competition between bulk energy away from the crack and surface energy on the crack. Contemporary fracture mechanics has completely espoused Griffith's viewpoint.

Of course, the post-Griffith development of fracture mechanics is paved with great contributions. Of special note is the link that G. R. Irwin provided between the bulk energy released during an infinitesimal advance of the crack (the energy release rate often denoted by G) and the coefficients weighing the singularity of the displacement field at the crack front (the stress intensity factors usually denoted by K_I, K_{II} or K_{III}). The celebrated Irwin's formulae – such as $G = K_{III}^2/2\mu$ in the anti-plane setting for linearized and isotropic elasticity – prompted and continue to prompt an avalanche of literature devoted to the computation of the stress intensity factors. This, we deem a mixed blessing because, while it is important to understand the detailed make-up of the elastic field near a singularity, it however drains expert energy away from the tenet of Griffith's approach: energetic competition.

We quote from (Griffith, 1920), p. 165–166, italicizing our additions to the quote: " In view of the inadequacy of the ordinary hypotheses, the problem of rupture of elastic solids has been attacked *by Griffith* from a new standpoint. According to the well-known "theorem of minimum energy", the equilibrium state of an elastic solid body, deformed by specified surface forces, is such that the potential energy of the whole system is a minimum. The new criterion of rupture is obtained by adding to this theorem the statement that the equilibrium position, if equilibrium is possible, must be one in which rupture of the solid has occurred, if the system can pass from the unbroken to the broken condition by a process involving a continuous decrease in potential energy.

In order, however, to apply this extended theorem to the problem of finding the breaking loads of real solids, it is necessary to take account of the increase in potential energy which occurs in the formation of new surfaces in the interior of such solids. ...... *For cracks extending several atomic lengths* the increase of energy, due to the spreading of the crack, will be given with sufficient accuracy by the product of the increment of surface into the surface tension of the material".

This tract decidedly adopts Griffith's variational viewpoint and heralds it as the foundation of fracture analysis. But the bias is rational and the proposed approach is a natural offspring of rational mechanics, as will be demonstrated hereafter. Actually, in its primal form, our take on fracture is completely equivalent to the "classical" viewpoint. The departure, when it occurs, will be a confession of mathematical inadequacy, rather than a belief in the soundness of additional physical principles, such as global minimality.

At this early stage, warnings to the reader should be explicit. At no point here do we contend that we have anything to contribute to "dynamic fracture". In the tradition of most works on fracture, kinetic effects are a priori assumed negligible and will remain so throughout. There will be no discussion of the restrictions on the loads that could validate such an assumption. Thus, the quasi-static hypothesis is the overarching non-negotiable feature. We do so after acknowledging that the intricacies created by hyperbolicity are not to be taken lightly. Here again, we confess our inadequacy but do not relish it. Quasi-statics view "time" as a synonym for a real ordered, positive parameter denoted by t and referred to as "time". All loads are functions of that parameter; an "evolution" corresponds to an interval of parametric values; "history" at time t is the remembrance of all parametric values $0 \leq s \leq t$.

Also, our tract is not to be construed as any kind of review of "classical" fracture. We make no effort to assess the existing literature and refrain from quoting even the most revered books on fracture. Our needs in that regard are modest and do not require to appeal to any of the highly sophisticated tools that have been developed since Griffith's seminal paper. For a general overview of those topics, the reader is directed to e.g. (Bui, 1978), (Leblond, 2000) for a more analytical presentation of the field, or to (Lawn, 1993) for a materials oriented view of fracture.

Without further ado, we now briefly provide a road-map for the study. The ideal reader is defined as follows: familiar with the basics of continuum mechanics, she will be accustomed with the rational mechanics formalization of thermo-mechanics in the sense of C. Truesdell. She will also demonstrate some familiarity with the classical minimality principles of linearized and hyper-elasticity, as well as some understanding of distributions, Sobolev spaces, the fundamentals of measure theory, integration and elliptic partial differential equations. Finally, it would be best if she also was somewhat versed in the classical theory of fracture mechanics, lest she should think that the expounded considerations are mere divagations. The actual reader is invited to consult the Appendix where she will find a self-contained, but succinct exposition of the necessary mathematical prerequisites. For this reason, we will not always refer to any particular text when using a "classical" result; the reader will find the corresponding statement in the Appendix. Although the initial mathematical investment is greatly reduced, in our opinion, by an attentive reading of the Appendix, we do not feel that we can provide an adequate exposition of the necessary mechanical prerequisites in a few pages and prefer to trust the reader's fluency in the basic tenet of elasticity, thermodynamics and fracture.

In a first section (Section 2), the classical theory is introduced within the rational framework. By classical theory, we mean that which Griffith introduced. In particular, the surface energy is the so-called Griffith's surface energy; its value is proportional to the crack area (crack length in 2d). The resulting problem is then re-formulated in a variational light. The end product may strike the mechanician as unfamiliar, yet not even the lighting is new if abiding by Griffith's previously quoted motto. With hindsight, we have just benefited from eighty years of mathematical experience since those lines were written. Actually, the analysis is performed for a larger class of energies, yet one that does not include cohesive type energies. We baptize those energies "Griffith-like".

As will be seen, the formulation is two-fold: an energy must remain stationary at every time among all virtual admissible crack-displacement pairs at that time, and an energy conservation statement must be satisfied throughout the time evolution. Stationarity statements are notoriously difficult to enforce without additional features such as local minimization. Our first and most egregious departure from the classical theory consists in replacing stationarity with local, or even worse, global minimality. This is the main price we are willing to pay in exchange for a meaningful theory. In the case of Griffith-like energies the notions of local and/or global minimality become notions of *unilateral* local and/or global minimality. As will become clearer in the sequel, the arguably labeled unilaterality refers to the unsightly presence of the local and/or global minimizer in the functional to be minimized. In other words, we will have to deal with minimization problems for a functional $\mathcal{P}$, where the unknown local and/or global minimizer u must minimize locally and/or globally $\mathcal{P}(u, v)$ for all admissible v's!

The subsidiary price tag is of a topological nature: the natural topology on cracks is unclear, especially in dimension 3, or when no connectedness is a priori assumed. It is then more convenient to view cracks as the compound location of all points where a displacement discontinuity has occurred throughout the history of the loading process. In other words, the crack-displacement pairs are replaced by displacements only, but those are in turn allowed to jump. The relevant ambient displacement space unfortunately allows for discontinuity sets that do not look like cracks at all, because their closure could be of positive Lebesgue measure. Most of the theoretical results are born out of this weak formulation, yet the regularity results that would permit one to recover a bona fide crack are almost non-existent. This should be mitigated by the riches brought about by the weak formulation; in the end, it is our belief that "the weak can overcome the strong"[1].

It would be ridiculous to a priori prohibit imports from any post-Griffith theory. In that respect the work of G.I. Barenblatt has particular significance because it pinpoints the potential deficiencies in Griffith's surface energy. First among those is the scaling effect that Griffith's surface energy imparts upon the formulation. Indeed that energy is proportional to surface area (or length in 2d), whereas the bulk energy varies like a volume (or an area in 2d). The ratio of bulk to surface energies is thus geometry dependent, and not only material dependent. So one should expect, for example, that the breaking pattern of a 1d-bar should be length-dependent, and this independently of the specific criterion adopted, provided that the surface energy is Griffith like (in this simplistic case a bang-bang $0, 1$ alternative). This is clearly nonphysical and can be remedied at once through the consideration of a cohesive type energy à la Barenblatt. We thus describe at the end of the first section the modifications to the formulation that accompany a Barenblatt type surface energy.

Section 3 is part of the discovery process in the stationarity vs. minimality litigation. The advocated departure from pure unilateral stationarity may ban sound evolutions and the proposed eugenic principle may even be so drastic that extinction of the evolutions will follow. Any kind of minimality principle in a non convex setting should raise suspicion; even more so in an evolutionary scheme. The section is but a timid intrusion into the debate in a minimalist environment: 1d traction, and anti-plane shear tearing. It is also designed to help the reader familiarize herself with the variational approach in settings where irreversibility – a delicate notion, see Section 5 — is automatically enforced. Both Griffith's and cohesive surface energies are considered and the existence of stationary evolutions, then locally minimizing evolutions, and finally globally minimizing evolutions is investigated. Of course any globally minimizing evolution is also a locally minimizing evolution and any locally minimizing evolution is also a stationary evolution. The tearing analytical experiment provides a stationary evolution which is unique, smooth, and also a locally minimizing and globally minimizing evolution. The case of 1d traction is much more intricate and a whole slew of evolutions is evidenced.

---

[1] Lao Tse – Tao Te King, 78

We would be ill-advised to draw general conclusions from those two examples and will only remark that well understood local minimality, together with a cohesive type energy, is very promising. The associated mathematical intricacies are a bit overwhelming at present, and much progress should be made before the proposed combination becomes a viable option in more complex situations, as will be seen through the remainder of this study.

Three main issues have plagued fracture mechanics in the last hundred years or so. Those are: initiation, irreversibility and path. By initiation, we mean nucleation of the crack, as well as further extension of the crack, given an observable pre-existing crack. Irreversibility is concerned with the definition of a threshold that marks the unrecoverable advance of the physical crack. Path encompasses all questions related to predictability of the geometric site of the future crack, given a loading histogram.

The next three sections of the paper revisit those issues in the light of the formulations developed in Section 2.

Section 4 tackles initiation through the variational prism. The choice of Griffith's surface energy yields too much, or too little. Too much when global minimality rules, because global minimizers become size-dependent – a straight manifestation of the already evoked scaling effect – and too little when any kind of local minimality criterion is activated, because generically the energy release rate is 0 and, consequently, no cracks will form.

A cohesive type energy fares much better, although the mathematical results are partial at best. Whenever local minimality – or even simply stationarity – presides, a critical yield stress (the slope of the surface energy at 0) determines the onset of the cracking process. If however global minimality is adopted, then a process zone will experience fine mixtures of large elastic deformations and small jumps. The resulting macroscopic behavior in the process zone will be plastic. The mechanical significance is portentous: cohesive brittleness leads to ductility!

In any case, the remarkable array of observed initiation patterns demonstrates the flexibility of the variational method: once a surface energy is picked, together with a minimality criterion, then the formulation delivers the initiation rule with no further ad-hoc import. The classically entertained notion of the original defect is no longer required.

Section 5 is concerned with irreversibility. In the case of a Griffith type energy, irreversibility is clear-cut. As already mentioned the crack will be the aggregate of the sites of all past jumps. We will show how this notion, easily implemented when the time evolution is made up of step increments in the loads, can be extended to a general setting where the loads vary arbitrarily with time. This will serve to illustrate the basic mathematical method in achieving existence results for quasi-static crack evolution. First the loads are discretized in time, and the ensuing step by step evolution is referred to as the incremental formulation, then the discretization step is sent to 0, yielding, in the best case scenario, the time-continuous formulation.

The case of a cohesive energy is more challenging because there is no obvious threshold for irreversibility. We will review possible choices, zeroing in on the correct choice if one strives to derive fatigue evolution from fracture evolution, an issue that will be further discussed in Section 9. As of yet, there is no analogue of the time-continuous evolution in the cohesive case and this is one of the mathematical challenges of the theory. We explain why this is so.

Section 6 is shorter and it investigates path. There, the only definitive results are those coming from the consideration of a global minimality criterion, together with a Griffith type surface energy. In that setting, path is a byproduct of the time-continuous evolution in the sense that, for a given solution to that evolution – one that satisfies global minimality at each time and energy conservation – there is a well-determined path that the crack will follow. The troubling issue is uniqueness, or apparent lack thereof. Indeed, there are at present no uniqueness results to speak of. In all fairness, this is not unlike many nonlinear, non-convex settings, such as buckling, where bifurcation branches are expected.

Many path-related issues are outstanding. Our pious wish to adjudicate the everlasting dispute between the $G_{\max}$–clan and the $K_{II} = 0$–clan, the two main opponents in the crack-branching conundrum remains exactly that. Our impotence in this respect is mitigated by numerics. The numerical treatment of the variational model is a vast topic, whose surface is barely scratched in Section 8. In any case, numerical evidence of branching is striking, as shown in Section 5, but numerics alone cannot provide the answer.

The remainder of the paper sits squarely within the confines of the variational method, because the survivors of the previous sections should by then be well acquainted with the main tenet of the approach.

Section 7 is our contribution to Griffith vs. Barenblatt. Within the variational framework the hypothetical convergence of cohesive type models to Griffith's model can be easily framed in the language of Γ–convergence, which we recall. Then, in the large domain limit, that is when the size of the investigated domain tends to infinity, the cohesive model with global minimality is shown to behave asymptotically like the Griffith's model with global minimality. We have unfortunately no significant contributions to put forth regarding this issue in the context of local minimality.

Section 8 is a peek through the numerical veil. As already stated, the topic is immense and our goal here is merely to provide the reader with a taste of the issues. That finite elements do not cope well with field discontinuities comes as no surprise. Consequently, the basic idea is to smear the discontinuities by adjoining an auxiliary field that will concentrate precisely around the discontinuities of the displacement field. The algorithm is then controlled by the thickness of the smear, which is assumed to be very small.

Mechanicians may be tempted to lend significance to the resulting model as a damage gradient model, like those developed in e.g. (Lorentz and Andrieux, 1999). For our part, we view it merely as an approximation and will resist any further discussion of its intrinsic physical merits.

The numerical study pertains to the global minimality setting, because it is the only one that allows detailed investigation, and also the only one for which a complete evolution has been derived. As a corollary, it only addresses Griffith type surface energies. The ensuing numerics are very stable and compliant, despite the lack of convexity of the two-field problem. Numerical illustrations are offered.

Section 9, the final section of this study, toys with cyclic loading, a.k.a fatigue. We demonstrate that, equipped with a Barenblatt type surface energy, an appropriate notion of irreversibility – that evoked in Section 4 – and global unilateral minimization, we can view fatigue, at least incrementally, in the same framework as fracture.

In 1d, the debonding of a thin film from a substrate provides a key to the derivation of fatigue debonding from cohesive fracture. Paris type laws are derived from fracture evolution, not a priori postulated.

Formidable mathematical hurdles prevent the consideration of more general settings, but, in our view, the seeds of the grand unification between fracture and fatigue are planted.

A few notational notes and/or cautionary notes follow in no particular order. Also consult the glossary for additional symbols used in the text.

Throughout, Einstein's summation convention is used.

The symbol $\mathcal{C}$, whenever it appears, refers to a generic positive and finite constant: for example, $2\mathcal{C}$ is replaced by $\mathcal{C}$.

The word iff stands for "if, and only if".

If $u : \Omega \mapsto \mathbb{R}^N$, then the linearized strain tensor $e(u) : \Omega \mapsto \mathbb{R}^{N^2}$ stands for $1/2(\nabla u + \nabla u^t)$.

The symbol $\lfloor$, applied to a set A, *i.e.*, $\lfloor A$, means "restricted to" A, while the symbol $\vee$ stands for "supremum", that is $a \vee b := \sup\{a, b\}$ for $a, b \in \mathbb{R}$.

The symbol $\#$ stands for "cardinal of".

For notational unity, we nearly always denote the kinematic field by φ. In most cases, φ should be thought of as the deformation field, but, on occasions, it will become the displacement field. This will be the case whenever anti-plane shear is at stake, unless the difference is made explicit as in Subsection 3.2. In this respect, homogeneity of the bulk energy in a 1d or 3d anti-plane shear setting will always refer to a property of the energy viewed as a function of the gradient of the displacement field. This should ease reader's angst, especially in Subsection 8.2 where homogeneity plays a crucial role in the "backtracking algorithm".

Dimensionality will not be set. Although simplicity of exposition dictates that the skeleton of the notes be two-dimensional, rather than three-dimensional, we will on occasion stumble into the third dimension or retreat to one dimension, when "confined in motion and eyesight to that single Straight Line"[2].

At the close of this introduction, we wish to dispel the notion that this study is a compendium of our contributions. Although no references were given as of yet, it should be self-evident that the sheer amount of results evoked above far exceeds our mutualized abilities. We will "render unto Caesar the things which are Caesar's"[3] throughout this paper. At this point we merely list, in alphabetical order, those who have and, for the most part, continue to contribute to the variational effort in the modeling of fracture: Jean-François Babadjian, Andrea Braides, François Bilteryst, Antonin Chambolle, Miguel Charlotte, Gianni Dal Maso, Gianpietro Del Piero, Alessandro Giacomini, André Jaubert, Christopher J. Larsen, Jérôme Laverne, Marcello Ponsiglione, Rodica Toader and Lev Truskinovsky.

---

[2] Edwin A. Abbott – Flatland
[3] Matthew – 22:21

2. Going variational

In this section, the starting premise is Griffith's model for crack evolution, as presented in his celebrated paper (Griffith, 1920). Of course, continuum mechanics has seen and weathered many storms in eighty seven years and it would make little sense to present fracture exactly as in (Griffith, 1920). The reader will find below what we believe to be a very classical introduction to brittle fracture within a rational mechanical framework. Whether this strictly conforms to the tenet of Rational Mechanics is a matter best left to the experts in the field.

Our starting assumptions are two-fold. First, as mentioned in the introduction, we do not wish to contribute at this point to the hesitant field of dynamic fracture, thereby restricting our focus to quasi-static evolution. At each time, the investigated sample is in static equilibrium with the loads that are applied to it at that time. We use the blanket label "loads" for both hard devices (displacement type boundary conditions) and soft devices (traction type boundary conditions and/or body forces). In the former case, we often refer to those boundary conditions as "displacement loads". Then, we do not concern ourselves with changes in temperature, implicitly assuming that those will not impact upon the mechanics of the evolution: in particular, thermal expansion is not an option in this model, at least to the extent that it couples thermal and mechanical effects. However, thermal stresses induced by a known temperature field fall squarely within the scope of the forthcoming analysis.

Also, we only discuss the 2d-case in this section. However, it will be clear that the resulting formulation applies as well to dimensions 1 and 3.

We consider Ω, a bounded open domain of $\mathbb{R}^2$. That domain is filled with a brittle elastic material. At this level of generality, the type of elastic behavior matters little, as long as it is represented by a bulk energy $F \mapsto W(F)$ which will be assumed to be a function of the gradient of the deformation field φ; in linearized elasticity W will become a function of $e(u) := \frac{1}{2}(\nabla u + \nabla u^t)$ with $\varphi(x) = x + u(x)$. We do not address invariance, objectivity, or material symmetry in the sequel, although isotropy will be a recurring feature of the many analytical and numerical examples discussed hereafter.

Time dependent loads are applied to Ω. We will assume that the force part of the load is given in the reference configuration (that is defined on $\overline{\Omega}$). Those are

- body forces denoted by $f_b(t)$ and defined on Ω;

- surface forces denoted by $f_s(t)$ and defined on $\partial_s \Omega \subset \partial \Omega$;

- boundary displacements denoted by $g(t)$ and defined on $\partial_d \Omega :=$ $\partial \Omega \backslash \partial_s \Omega$. Precisely, we assume throughout that $g(t)$ is defined and smooth enough on all of $\mathbb{R}^2$ and that the boundary displacement is the trace of $g(t)$ on $\partial_d \Omega$.

The backdrop is in place and Griffith may now enter the stage.

2.1. GRIFFITH'S THEORY

The theory espoused by Griffith is macroscopic in scope and mechanical in essence. The crack or cracks are geometrically idealized as discontinuity surfaces for the deformation field of the continuum under investigation. If that continuum behaves elastically, material response under external loading will be unambiguous once the laws that preside over the onset and propagation of the crack(s) are specified. The construction of such laws – the goal of Griffith's theory – requires three foundational ingredients,

1. A surface energy associated with the surfaces where the deformation is discontinuous;

2. A propagation criterion for those surfaces;

3. An irreversibility condition for the cracking process.

The surface energy adopted by Griffith is simple, even simplistic in the eyes of the post-modern solid state physicist. Throughout the cracking process, a(n isotropic) homogeneous material spends an energy amount which remains proportional to the area of the surface of discontinuity. We take license to call fracture toughness of the material the proportionality factor, and denote it by k, while being aware that fracture toughness habitually refers to the mode-I critical stress intensity factor in isotropic linearized elasticity.

As already noted in the introduction, Griffith eagerly confesses in (Griffith, 1920) to the limits of validity of that energy. Griffith's energy is the macroscopic manifestation of the energy spent through the microscopic breaking of inter-atomic bonds. A simple counting argument demonstrates that, if inter-atomic bonding is ruled by a Lennard-Jones type interaction potential, then the add-energy spent in moving two atoms apart while the remaining atoms stay put is additive, which ultimately yields a total (macroscopic) energy proportional to the separation area. Thus, for Griffith's energy to apply the break-up must be final. In macroscopic words, the jump in displacement on the crack site must have exceeded some threshold. In the absence of contact the crack lips do not interact and cohesiveness is prohibited.

We observed in the introduction that the propagation criterion is energy based. The test is a balance between the potential energy released through a virtual increase of the crack length (area) and the energy spent in creating additional length (area). The crack will extend only if the balance favors creation.

It will be seen later, most notably in Sections 5 and 9, that irreversibility is not a straightforward concept in the presence of cohesiveness. However Griffith's energy presupposes the absence of cohesive forces. Thus a crack will form where and at the time at which the displacement field becomes discontinuous. It will then stay so "in saecula saeculorum", oblivious to the actual state of displacement at any posterior time. We emphasize that the approach advocated in this tract treats cohesive forces as a simple byproduct of the surface energy; see Subsection 5.2. In other words, the presence of such forces is conditioned by the proper choice of surface energy.

We now formulate Griffith's view of the crack evolution problem in a(n isotropic) homogeneous elastic material.

For now, the crack path $\hat{\Gamma}$ is assumed to be known *a priori*. We wish to include partial debonding as a possible crack behavior, so that $\hat{\Gamma} \subset \overline{\Omega}\backslash\partial_s\Omega$. The crack at time t is assumed to be a time increasing connected subset of $\hat{\Gamma}$; it can thus live partially, or totally on $\partial\Omega$. It is therefore completely determined by its length l and denoted by $\Gamma(l)$.

By the quasi-static assumption, the cracked solid (see Figure 2.1) is, at each time, in elastic equilibrium with the loads that it supports at that time; in other words, if the crack length at that time is l, then the kinematic unknown at that time, $\varphi(t, l)$ (the transformation, or displacement) satisfies

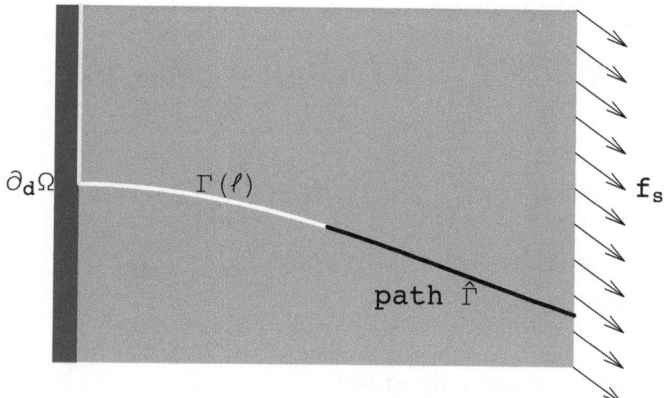

Figure 2.1. The cracked solid

$$\begin{cases} -\mathrm{div}\,\dfrac{\partial W}{\partial F}(\nabla\varphi(t,l)) = f_b(t) \text{ in } \Omega\backslash\Gamma(l) \\[2mm] \varphi(t,l) = g(t) \text{ on } \partial_d\Omega\backslash\Gamma(l) \\[2mm] \dfrac{\partial W}{\partial F}(\nabla\varphi(t,l))n = f_s(t) \text{ on } \partial_s\Omega \\[2mm] \dfrac{\partial W}{\partial F}(\nabla\varphi(t,l))n = 0 \quad \text{on } \overline{\Omega}\cap\Gamma(l) \end{cases} \tag{2.1}$$

where n denotes the appropriate normal vector.

The last relation in (2.1) calls for several comments. In an anti-plane shear setting, it merely states, in accord with Griffith's premise, the absence of cohesive forces along the crack lips. In a planar situation, it implicitly assumes separation of the crack lips, hence non-interpenetration. In all honesty, we will systematically skirt the issue of non-interpenetration in our presentation of Griffith's evolution; in the geometrically non-linear setting of hyperelasticity, it is an issue even in the absence of brittleness (Ciarlet and Nečas, 1987). Implementation of a condition of non-interpenetration at the crack lips, be it in the non-linear or in the linearized context, raises multiple issues that go beyond the scope of this review. It is also our admittedly subjective view that non-interpenetration matters little when trying to capture the main features of crack propagation in a Griffith setting. By contrast, non-interpenetration is, as will be seen later, an essential feature of cohesive models and we squarely confront the issue in that setting (see Subsections 4.2, 5.2).

The system (2.1) assumes that the crack length is known. Griffith's decisive input is to propose the following criteria for the determination of that length. At time t, compute the potential energy associated with the crack of length l, that is

$$\mathcal{P}(t,l) := \int_{\Omega\backslash\Gamma(l)} W(\nabla\varphi(t,l))\,dx - \mathcal{F}(t,\varphi(t,l)) \tag{2.2}$$

with

$$\mathcal{F}(t,\varphi) := \int_{\Omega} f_b(t)\cdot\varphi\,dx + \int_{\partial_s\Omega} f_s(t)\cdot\varphi\,ds. \tag{2.3}$$

Then, $l(t)$ must be such that it obeys

- The Griffith's criterion:

a. $l \nearrow^t$ (the crack can only grow);

b. $-\partial\mathcal{P}/\partial l(t,l(t)) \leq k$ (the energy release rate is bounded from above by the fracture toughness);

c. $(\partial\mathcal{P}/\partial l(t, l(t)) + k)\,\dot{l}(t) = 0$ (the crack will not grow unless the energy release rate is critical).

From a thermodynamical viewpoint, Griffith's criterion should be interpreted as follows. The crack length is a global internal variable, and its variation induces a dissipation which must in turn satisfy Clausius–Duhem's inequality.

The attentive reader will object that the definition of the potential energy becomes specious as soon as the kinematic field $\varphi(t, l)$ fails to be uniquely defined – as is for instance the case in hyperelasticity – and that, consequently, Griffith's criterion is meaningless in such a setting. We readily concede and remark that the very definition of a global internal variable in the absence of convexity challenges classical thermomechanics. The reader is thus invited to assume the existence of a solution path $\varphi(t, l(t))$ for which the associated potential energy $\mathcal{P}(t, l(t))$ is well-behaved, lending meaning to the Griffith's criterion. Section 5, and particularly Theorem 5.5 will hopefully demonstrate that the assumption is not totally without merit, at least when filtered through the approach proposed in this work.

A convenient enforcement of Clausius–Duhem's inequality is provided through the introduction of a convex dissipation potential $\mathcal{D}(\dot{l})$, further satisfying $\mathcal{D}(0) = 0$. Then, the inequality reduces to

$$-\frac{\partial\mathcal{P}}{\partial l}(t, l(t)) \in \partial\mathcal{D}(\dot{l}(t)). \tag{2.4}$$

The correct dissipation potential in Griffith's setting is denoted by $\mathcal{D}_G$ and given by (see Figure 2.2)

$$\mathcal{D}_G(\dot{l}) := \begin{cases} k\dot{l}, & \dot{l} \geq 0 \\ \infty, & \dot{l} < 0, \end{cases} \tag{2.5}$$

and (2.4) then yields precisely Griffith's criteria. So, summing up, Griffith's modeling of crack evolution reduces to (2.1), (2.4) with (2.5) as dissipation potential.

As we will see, positive 1-homogeneity is the vital feature of the dissipation potential, if one is to adopt a variational viewpoint and hope for a time-continuous evolution. It will become handy in Section 9 to consider potentials for which 1-homogeneity does not hold. We refer to Subsection 2.4 below in that case.

From the thermodynamical standpoint, Griffith's dissipation potential can be greatly generalized. The crack may be thought of as depending on several global internal variables, say $(l_1, ..., l_p)$. In other words the value of the p-uple $l := (l_1, ..., l_p)$ determines the crack length now denoted by $\ell(l)$, hence the crack itself, which is still denoted

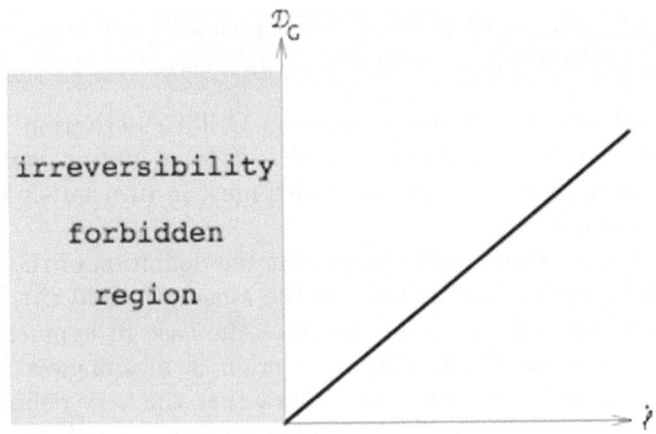

Figure 2.2. Griffith dissipation potential

by $\Gamma(l)$. Then $\mathcal{D}_G$ can be replaced by any positive convex potential prohibiting crack decrease, that is that (2.4) can be applied to any "Griffith-like" potential of the form

$$\mathcal{D}(l;\dot{l}) := \begin{cases} D(\dot{l}), & \text{if } \nabla\ell(l)\cdot\dot{l}\geq 0 \\ \infty, & \text{otherwise} \end{cases} \tag{2.6}$$

with $D : \mathbb{R}^p \mapsto \mathbb{R}^+$ convex, $D(0) = 0$.

We are now ready to explore the system (2.1), (2.4) with (2.6) as dissipation potential and $\nabla_l\mathcal{P}$ replacing $\partial\mathcal{P}/\partial l$. For completeness, we should add an initial condition to (2.4); we will thus assume that

$$l(0) = l_0, \tag{2.7}$$

and denote, from now onward, any pair-solution $(l(t), \varphi(t, l(t)))$, if it exists, by $(l(t), \varphi(t))$.

2.2. THE 1-HOMOGENEOUS CASE – A VARIATIONAL EQUIVALENCE

Throughout this subsection, we assume that the Griffith-like potential is positively 1-homogeneous, which amounts to a statement of rate-independence, as explained at length in various works; see (Mielke, 2005) for a general treatment of rate independent processes, and also (Francfort and Mielke, 2006). Rate independence is clearly a feature of quasi-static crack evolution within the framework developed by Griffith.

Assuming suitable – and unstated – smoothness of all relevant quantities, we propose to establish the equivalence between the original system (2.1), (2.4), (2.7) and a two-pronged formulation which states that a certain energy must remain stationary at every time among all virtual admissible crack-displacement pairs at that time, and that an energy conservation statement must be satisfied throughout the time evolution. This is the object of the following

PROPOSITION 2.1. *Assuming that the potential D in (2.6) is positively 1-homogeneous, then the pair $(l(t), \varphi(t))$ (satisfying (2.7)) satisfies (2.1), (2.4) (with appropriate smoothness) on $[0, T]$ iff, for every $t \in [0, T]$, it satisfies (with that same smoothness)*

(Ust) $(l(t), \varphi(t))$ *is a stationary point of*

$$\mathcal{E}(t; \varphi, l) := \int_{\Omega \backslash \Gamma(l)} W(\nabla \varphi) \, dx - \mathcal{F}(t, \varphi) + \mathcal{D}(l(t); l - l(t)), \quad (2.8)$$

among all l and all $\varphi = g(t)$ on $\partial_d \Omega \backslash \Gamma(l)$ in the sense of (2.11) below;

(Ir) $\dot{\ell}(t) = \nabla \ell(l(t)) \cdot \dot{l}(t) \geq 0;$

(Eb) $\dfrac{dE}{dt}(t) = \int_{\partial_d \Omega \backslash \Gamma(l(t))} \dfrac{\partial W}{\partial F}(\nabla \varphi(t)) n \cdot \dot{g}(t) \, ds - \dot{\mathcal{F}}(t, \varphi(t))$

with

$$\dot{\mathcal{F}}(t, \varphi) := \int_\Omega \dot{f}_b(t) \cdot \varphi \, dx + \int_{\partial_s \Omega} \dot{f}_s(t) \cdot \varphi \, ds \qquad (2.9)$$

$$E(t) := \int_{\Omega \backslash \Gamma(l(t))} W(\nabla \varphi(t)) \, dx - \mathcal{F}(t, \varphi(t)) + \int_0^t D(\dot{l}(\tau)) d\tau$$

$$= \mathcal{P}(t, l(t)) + \int_0^t D(\dot{l}(\tau)) d\tau. \qquad (2.10)$$

The unilateral stationarity statement (Ust) is rather unusual because the functional $\mathcal{E}(t; \cdot)$ that should be stationary at $(l(t), \varphi(t))$ explicitly depends on $l(t)$; hence the label unilateral. The energy balance (Eb) can be turned, through various integration by parts in time, into what is referred to in the literature as the mechanical form of the second law of thermodynamics; see e.g. (Gurtin, 2000).

Proof. First we should clearly articulate what is meant by (Ust). To this effect, we introduce a one-parameter family of variations of the kinematic variable $\varphi(t)$ and of the crack length $l(t)$ as follows. We set

$$l(t, \varepsilon) := l(t) + \varepsilon \hat{l} \, ; \quad \varphi(t, \varepsilon, l) := \varphi(t, l) + \varepsilon \psi(t, l),$$

where $\psi(t, l) = 0$ on $\partial_d\Omega\backslash\Gamma(l)$ and $\varphi(t, l(t)) = \varphi(t)$. Then, unilateral stationarity is meant as

$$\frac{d}{d\varepsilon}\mathcal{E}(t; \varphi(t, \varepsilon, l(t, \varepsilon)), l(t, \varepsilon))\Big|_{\varepsilon=0} \geq 0. \tag{2.11}$$

Recall the expression (2.8) for $\mathcal{E}$, and use positive 1-homogeneity, so that $\mathcal{D}(l(t); \varepsilon l) = \varepsilon\mathcal{D}(l(t); l)$. Then, the above also reads as

$$\int_{\Omega\backslash\Gamma(l(t))} \frac{\partial W}{\partial F}(\nabla\varphi(t)) \cdot \nabla\psi \, dx - \mathcal{F}(t, \psi) + \nabla_l P(t, l(t)) \cdot \hat{l} + \mathcal{D}(l(t); \hat{l}) \geq 0,$$

where we recall that $\mathcal{P}$ was defined in (2.2). Consequently, through integration by parts, (Ust) is equivalent to

$$(2.1) \quad \text{and} \quad \nabla_l P(t, l(t)) \cdot \hat{l} + \mathcal{D}(l(t); \hat{l}) \geq 0, \; \forall \hat{l}. \tag{2.12}$$

Then, assume that (Ust), (Ir), (Eb) hold. In view of the above, (2.1) is satisfied, so that (Eb) reduces to

$$\nabla_l P(t, l(t)) \cdot \dot{l}(t) + \mathcal{D}(\dot{l}(t)) = 0. \tag{2.13}$$

Subtracting (2.13) from (2.12), we conclude, with (Ir), that

$$-\nabla_l P(t, l(t)) \in \partial\mathcal{D}(l(t); \dot{l}(t)),$$

which is precisely (2.4) (the sub-differential being evaluated with respect to the second variable $\dot{l}$).

Conversely, if (2.1) holds true, then

$$\frac{dE}{dt}(t) = \int_{\partial_d\Omega\backslash\Gamma(l(t))} \frac{\partial W}{\partial F}(\nabla\varphi(t)) \cdot \dot{g}(t) \, ds - \dot{\mathcal{F}}(t, \varphi(t))$$
$$+ \left\{\nabla_l P(t, l(t)) \cdot \dot{l}(t) + \mathcal{D}(\dot{l}(t))\right\}. \tag{2.14}$$

But, by 1-homogeneity, $\mathcal{D}(\dot{l}(t)) = \dot{l}(t)\zeta, \forall\zeta \in \partial\mathcal{D}(l(t); \dot{l}(t))$, so that, if (2.4) also holds, then the term in brackets in (2.14) cancels out and (Eb) is established. In view of (2.12), it remains to show that

$$\nabla_l P(t, l(t)) \cdot \hat{l} + \mathcal{D}(l(t); \hat{l}) \geq 0, \; \forall \hat{l}.$$

From (2.4), we get, since $l(t), \dot{l}(t)$ satisfy (Ir) by the definition (2.6) of $\mathcal{D}$,

$$-\nabla_l P(t, l(t)) \cdot \lambda\hat{l} \leq \mathcal{D}(\dot{l}(t) + \lambda\hat{l}) - \mathcal{D}(\dot{l}(t)), \; \lambda \geq 0.$$

Dividing by λ, using 1-homogeneity and letting λ tend to ∞, we recover the inequality in (2.12) since $\mathcal{D}(0) = 0$. Hence (Ust). □

REMARK 2.2. In the strict Griffith setting, the expressions for (Ust) and (Eb) simplify a bit, since $\mathcal{D}$ is linear on $\mathbb{R}^+$; in particular, since $\mathcal{D}(l - l(t)) = k(l - l(t))$ as soon as $l \geq l(t)$, the explicit dependence of $\mathcal{E}(t; \cdot)$ upon $l(t)$ drops out of that expression; it is still however a constraint on the admissible lengths. We rewrite (Ust), (Ir) and (Eb) below for the reader's convenience:

(Ust) $(l(t), \varphi(t))$ is a stationary point of

$$\mathcal{E}(t; \varphi, l) := \int_{\Omega \backslash \Gamma(l)} W(\nabla \varphi) \, dx - \mathcal{F}(t, \varphi) + kl, \qquad (2.15)$$

among all $l \geq l(t)$ and all $\varphi = g(t)$ on $\partial_d \Omega \backslash \Gamma(l)$;

(Ir) $\dot{l}(t) \geq 0$;

(Eb) $\dfrac{dE}{dt}(t) = \displaystyle\int_{\partial_d \Omega \backslash \Gamma(l(t))} \dfrac{\partial W}{\partial F}(\nabla \varphi(t)) \cdot \dot{g}(t) \, ds - \dot{\mathcal{F}}(t, \varphi(t))$

with

$$\begin{aligned} E(t) &:= \int_{\Omega \backslash \Gamma(l(t))} W(\nabla \varphi(t)) \, dx - \mathcal{F}(t, \varphi(t)) + kl(t) \\ &= \mathcal{P}(t, l(t)) + kl(t). \end{aligned} \qquad (2.16)$$

Throughout most of the remainder of this study, (Ust) and (Eb) will refer to the expressions above.

REMARK 2.3. In the context of Remark 2.2, elimination of the kinematic field in the variational formulation leads to the sometimes more convenient equivalent formulation for (Ust) below; however, the reader is reminded to keep in mind that lack of convexity, or rather of uniqueness challenges the very meaning of the potential energy.

(Ust) $l(t)$ is a stationary point of $\mathcal{P}(t, l) + kl$, among all $l \geq l(t)$.

At this point, we wish to strongly emphasize that, modulo smoothness, Griffith's formulation and the variational formulation obtained in Proposition 2.1 and in Remark 2.2 are strictly *one and the same* and cannot be opposed on mechanical grounds anymore than the original formulation. Of course, ill-wishers might object to the smoothness caveat, but pre-assuming smoothness is universal practice in deriving a notion of weak solution, so that we feel perfectly justified in doing so, and will be quite ready to qualify as "weak" the solutions of what we will, from now onward, label the "variational evolution".

In any case, Griffith's formulation is pregnant with smoothness-related issues as demonstrated in the next subsection.

2.3. SMOOTHNESS – THE SOFT BELLY OF GRIFFITH'S FORMULATION

Consider the case of a $p > 1$-homogeneous elastic energy density and of a monotonically increasing load, that is

$$W(tF) = t^p W(F), \qquad \mathcal{F}(t, t\varphi) = t^p \mathcal{F}(1, \varphi), \quad g(t) = t\,g.$$

Then, by homogeneity,

$$\varphi(t, l) = t\bar{\varphi}(l), \qquad \mathcal{P}(t, l) = t^p \bar{\mathcal{P}}(l),$$

where $\bar{\varphi}(l)$ and $\bar{\mathcal{P}}(l)$ are respectively the transformation and the potential energy associated with a crack of length l and loads corresponding to the value $t = 1$. Truly, from a mechanics standpoint, the displacement field $u(x) = \varphi(x) - x$ is the kinematic variable for which p-homogeneity of the associated energies makes sense. The conclusions drawn below would remain unchanged in that context.

We assume that $\bar{\mathcal{P}}$ is a sufficiently smooth function of l and focus on the initiation of an add-crack, starting with a crack of length l_0. Then, if $\bar{\mathcal{P}}'(l_0) = 0$, the crack will never move forward, so that $l(t) = l_0, \forall t$, whereas if $\bar{\mathcal{P}}'(l_0) < 0$, the crack will start propagating for

$$t_0 := \sqrt[p]{\frac{k}{-\bar{\mathcal{P}}'(l_0)}}.$$

As will be seen in Section 4, the energy release rate will be 0 for $l = l_0$, unless the elastic field happens to be sufficiently singular at the initiation point (this is the notion of "not-weak" singularity). In particular, a crack-free sample with a nice boundary will never undergo crack initiation, as usually professed in the fracture community.

Assuming thus that $\bar{\mathcal{P}}'(l_0) < 0$, we proceed to investigate the convexity properties of $\bar{\mathcal{P}}$ at l_0. If $\bar{\mathcal{P}}''(l_0) < 0$, then $\bar{\mathcal{P}}'$ is a strictly monotonically decreasing function in a neighborhood of l_0. Any smooth evolution $l(t)$ will then violate Griffith's criterion because

$$-t^p \bar{\mathcal{P}}'(l(t)) > -t_0^p \bar{\mathcal{P}}'(l_0) = k,$$

for t slightly larger than t_0.

In fact, in the restricted context of this subsection, the strict convexity of $\bar{\mathcal{P}}$ is a necessary and sufficient condition for the existence of a unique smooth crack evolution, as demonstrated by the following

PROPOSITION 2.4. *Given a smooth potential energy $\bar{\mathcal{P}}$, that energy is a strictly convex function of l on $[l_0, l_1]$, iff Griffith's criterion is satisfied by a unique smooth crack propagation $l(t)$ on $[t_0, t_1]$ given by*

$$l(t) = (\bar{\mathcal{P}}')^{-1}\left(-\frac{k}{t^p}\right), \qquad t_1 = \sqrt[p]{\frac{k}{-\bar{\mathcal{P}}'(l_1)}}. \qquad (2.17)$$

Then, at each time t, $-t^p \bar{\mathcal{P}}'(l(t)) = k$.

Proof. If $\bar{\mathcal{P}}$ is strictly convex, then (2.17) is well defined, smooth, and clearly satisfies Griffith's criterion. A solution $\tilde{l}(t) \not\equiv l(t)$ must be such that $-t^p \bar{\mathcal{P}}'(\tilde{l}(t)) < k$ on some sub-interval $(a, b) \subset [l_0, l_1]$. Then $\tilde{l}(t) = l(a)$ for $a \le t \le b$, hence $k = -a^p \bar{\mathcal{P}}'(\tilde{l}(a)) < -t^p \bar{\mathcal{P}}'(\tilde{l}(a)) = -t^p \bar{\mathcal{P}}'(\tilde{l}(t)) < k$ on (a, b), which is impossible.

Conversely, if a smooth function $l(t)$ is the only one that satisfies Griffith's criterion on $[t_0, t_1]$, $t_0 < t_1$, then, if $l(t_1) = l_0$, there is nothing to prove. Otherwise, let l and l_* be such that $l_0 < l < l_* < l(t_1)$. Those lengths are attained on time intervals $[t, t']$ and $[t_*, t'_*]$ with $t \le t'$, $t' < t_*$, $t_* \le t'_*$. Further, $-(t')^p \bar{\mathcal{P}}'(l(t')) = -(t'_*)^p \bar{\mathcal{P}}'(l(t'_*)) = k$, so that

$$\bar{\mathcal{P}}'(l) = -\frac{k}{(t')^p} < -\frac{k}{(t'_*)^p} = \bar{\mathcal{P}}'(l_*),$$

hence the strict convexity of $\bar{\mathcal{P}}$. □

This simple proposition has striking consequences. It demonstrates, albeit in a restrictive setting, that smoothness of the propagation inevitably leads to a reinforcement of the unilateral stationarity principle (Ust). The crack length $l(t)$ must actually be a minimizer for $\mathcal{P}(t, l) + kl$, because of the necessary convexity of $\mathcal{P}$.

So Griffith's criterion, which is ab initio non-sensical for non-smooth crack evolutions, implicitly pre-supposes the global convexity of the potential energy as a function of the crack length. "The intimacy of a well-kept secret"[4] is unraveled.

As mentioned in the introduction, stationarity is not a very pleasant mathematical notion from the standpoint of existence and it it is tempting to somewhat strengthen (Ust). Observe that (Ust) amounts to a first order optimality condition for $(l(t), \varphi(t))$ to be a local unilateral minimizer – in any reasonable topology – of $\mathcal{E}(t; \cdot)$.

The preceding analysis strongly militates for the adoption of some kind of minimality principle. Consequently, we propose the following two levels of departure from Griffith's classical theory:

– Local level – (Ust) is replaced by (Ulm) $(l(t), \varphi(t))$ is a local minimizer (in a topology that remains to be specified) for $\mathcal{E}(t; \varphi, l)$ among all $l \ge l(t)$ and all $\varphi = g(t)$ on $\partial_d \Omega \backslash \Gamma(l)$;

---

[4] Marguerite Yourcenar - L'Œuvre au Noir

– Global level – (Ust) is replaced by (Ugm) $(l(t), \varphi(t))$ is a global minimizer for $\mathcal{E}(t; \varphi, l)$ among all $l \geq l(t)$ and all $\varphi = g(t)$ on $\partial_d \Omega \backslash \Gamma(l)$.

In so doing, we have in effect selected solution-paths. The use of local minimality of the energy functional as a selection criterion is common practice for non linear conservative systems exhibiting a lack of uniqueness. The argument finds its "raison d'être" in the rigorous equivalence between Lyapunov stability and local minimality for systems with a finite number of degrees of freedom. Even more to the point, the search for global minimizers of the potential energy in finite elasticity has largely overshadowed that of stationary points. Of course, in our setting, the minimization criterion, be it global or local, must also accommodate irreversibility, hence the already mentioned notion of *unilaterality*.

In any case, Newton's law and Cauchy's theorem do not and will never imply minimality and the adopted minimality principles should be seen as postulates. Similar criteria have proved successful for many a dissipative system – see e.g. (Nguyen, 2000) – and we merely extend them to the setting of fracture. In that setting, it would be presumptuous to assume that any kind of local minimality statement can be derived from some undefined evolution, the more so because the model is already time-dependent.

REMARK 2.5. In all fairness, the exploration of brittle fracture as the asymptotic state of a dissipative system can be fathomed within the framework of quasi-static visco-elasticity. The viscosity is the vanishing parameter. The resulting system is not conservative, in the sense that energy balance should no longer hold true, some amount of energy – on top of the surface energy – being dissipated in the zero-viscosity limit. If successful, that route would lead to locally minimizing energy paths with a possible decrease in the energy at a point of discontinuity for the crack length or for the crack path.

The implementation of such a scheme is not straightforward. A first attempt may be found in (Toader and Zanini, 2005) in a two-dimensional setting. The crack path is prescribed and the crack is assumed to be connected. A locally minimizing possibly dissipative path is then generated, and is proposed as a potential competitor against the globally minimizing path. The verdict is postponed, pending further investigation.

Our approach will be more pragmatic. We will carefully dissect the consequences of those minimality principles and attempt to give a nuanced account of their respective merits.

Before we proceed, we would like to point out that, in the absence of 1-homogeneity, one can still hope for some kind of variational evolution. The argument will not be pristine, and will be reminiscent of the rate formulations common in plasticity. This is the object of the very short Subsection 2.4. Once this is done, we will return to the time-continuous variational evolution and recast it in a more suitable functional framework in Subsection 2.5.

2.4. THE NON 1-HOMOGENEOUS CASE – A DISCRETE VARIATIONAL EVOLUTION

As we saw previously, the 1-homogeneous character of the dissipation potential played a pivotal role in the derivation of (Ust), (Ir), (Eb). Absent this restriction, we cannot hope to prove any kind of equivalence. Any attempt at numerically solving (2.1),(2.4),(2.7) would certainly take its root in a time-stepping procedure. We propose now to travel a bit along that path. To that end, we consider a partition $0 = t_0 < ..... < t_i^n < ... < t_n^n = T$ of $[0, T]$ with $\Delta_n = t_{i+1}^n - t_i^n$.

Finite-differencing (Ust), (Ir), (Eb) (for general convex dissipation potentials $\mathcal{D}$ of the form (2.6)), we obtain, together with (2.7),

$$
\begin{cases}
-\text{div}\, \dfrac{\partial W}{\partial F}(\nabla \varphi_{i+1}^n) = f_b(t_{i+1}^n) \text{ in } \Omega \backslash \Gamma(l_{i+1}^n) \\[2mm]
\varphi_{i+1}^n = g(t_{i+1}^n) \text{ on } \partial_d \Omega \backslash \Gamma(l_{i+1}^n) \\[2mm]
\dfrac{\partial W}{\partial F}(\nabla \varphi_{i+1}^n)n = f_s(t_{i+1}^n) \text{ on } \partial_s \Omega \\[2mm]
\dfrac{\partial W}{\partial F}(\nabla \varphi_{i+1}^n)n = 0 \quad \text{on } \overline{\Omega} \cap \Gamma(l_{i+1}^n),
\end{cases}
\tag{2.18}
$$

and

$$
-\nabla_l \mathcal{P}(t_{i+1}^n, l_{i+1}^n) \in \partial \mathcal{D}(l_i^n; \frac{l_{i+1}^n - l_i^n}{\Delta_n}).
\tag{2.19}
$$

In turn, a pair solution $(l_{i+1}^n, \varphi_{i+1}^n)$ of the above system with $\nabla \ell(l_i^n)$ $(l_{i+1}^n - l_i^n) \geq 0$ may easily be seen to be a unilateral stationary point of

$$
\mathcal{E}_{i+1}^n(v, l) := \int_{\Omega \backslash \Gamma(l)} W(\nabla \varphi)\, dx - \mathcal{F}(t_{i+1}^n, \varphi) + \Delta_n \mathcal{D}(l_i^n; \frac{l - l_i^n}{\Delta_n}),
$$

among all $\nabla \ell(l_i^n)(l - l_i^n) \geq 0$, $\varphi = g(t_{i+1}^n)$ on $\partial_d \Omega \backslash \Gamma(l)$, and conversely. In the case where $\mathcal{D}$ is positively 1-homogeneous the Δ_n cancel out and we recover a discretized version of (Ust). Note that (Eb) seems to have dropped out of the discrete formulation altogether. We will come back to this point in Section 5.

The next natural step would be to pass to the limit in the discrete variational evolution as $\Delta_n \searrow 0$. This is unfortunately a formidable task, even in much simpler settings. For example, there are at present no mathematical results permitting to pass to the limit in the discrete gradient flow problem for a non-convex functional. In other words, if $W : F \in \mathbb{R}^{d \times d} \mapsto \mathbb{R}$ is a typical hyperelastic energy – say a convex function of the minors of F – then, under appropriate growth and boundary conditions, the global minimization problem

$$\min_{\varphi} \left\{ \int_\Omega W(\nabla \varphi) \, dx + \frac{1}{2\Delta_n} \int_\Omega (\varphi - \varphi_i^n)^2 \, dx \right\}$$

admits a solution with $\varphi_0^n = \varphi_0$ prescribed, and a priori estimates can be obtained on $\varphi^n(t)$, constructed as the piecewise constant function $\varphi^n(t) := \varphi_i^n$, $t \in [t_i^n, t_{i+1}^n)$. Yet what the appropriate weak limit of φ^n satisfies is unclear. In other words, there is no well-formulated L^2-gradient flow for non-convex functionals.

    In Section 9, we will make use of the procedure described above to propose a time-discretized variational evolution for fatigue. For now, we return to our main concern, the variational evolution described in Remark 2.2, and propose a functional framework that makes its analysis more palatable.

2.5. FUNCTIONAL FRAMEWORK – A WEAK VARIATIONAL EVOLUTION

The end of Subsection 2.3 emphasized the drawbacks of replacing (Ust) by (Ulm), or, even worse, by (Ugm). But the strengthened formulation "makes light out of darkness"[5], because, thanks to the minimality criterion, the preset path constraint can be abolished. Indeed, the minimality-modified Griffith variational evolution states that the actual length $l(t)$ of the crack is a local (or global) minimum among all lengths l greater than, or equal to $l(t)$ along the pre-determined crack path $\hat{\Gamma}$. But, why should one restrict the future evolution precisely to that curve $\hat{\Gamma}$? We thus propose to let the crack choose which future path it wishes to borrow, according to the minimality principle. Thus, denoting by $\Gamma(t)$ the crack at time t, we replace (Ulm), resp. (Ugm) by

(Ulm) $(\Gamma(t), \varphi(t))$ is a local minimizer (in a topology that remains to be specified) for

$$\mathcal{E}(t; \varphi, \Gamma) := \int_{\Omega \backslash \Gamma} W(\nabla \varphi) \, dx - \mathcal{F}(t, \varphi) + k\mathcal{H}^1(\Gamma), \qquad (2.20)$$

among all $\Gamma \supset \Gamma(t)$ and all $\varphi = g(t)$ on $\partial_d \Omega \backslash \Gamma$; or, resp.,

---

[5] Job - 37:15

(**Ugm**) $(\Gamma(t), \varphi(t))$ is a global minimizer for $\mathcal{E}(t; \varphi, \Gamma)$ among all $\Gamma \supset \Gamma(t)$ and all $\varphi = g(t)$ on $\partial_d \Omega \backslash \Gamma$.

Note that the test φ's depend on the test Γ's. Correspondingly, we also replace (2.7) by

(**Ic**) $\Gamma(0) = \Gamma_0$,

and the definition (2.16) of $E(t)$ in (Eb) by

$$
\begin{aligned}
E(t) &:= \int_{\Omega \backslash \Gamma(t)} W(\nabla \varphi(t)) \, dx - \mathcal{F}(t, \varphi(t)) + k \mathcal{H}^1(\Gamma(t)) \\
&= \mathcal{P}(t, \Gamma(t)) + k \mathcal{H}^1(\Gamma(t)),
\end{aligned}
\tag{2.21}
$$

with an obvious extension of the definition (2.2) of the potential energy $\mathcal{P}$.

This calls for two remarks. First, we keep the same label for those extended minimality principles, because they will be the only ones we will refer to from now onward. Then, we allow the test cracks Γ to be pretty much any *closed* set in $\overline{\Omega} \backslash \partial_s \Omega$ with finite Hausdorff measure $\mathcal{H}^1(\Gamma)$. This allows us to envision very rough cracks, and will coincide with the usual length when the crack is a rectifiable curve. We do not allow for the crack to lie on $\partial_s \Omega$ for obvious reasons. The crack cannot live where soft devices are applied, lest those soft devices not be felt.

We shall refer to the above formulation, that is (Ic), (Ulm) or (Ugm), (Eb), as *the strong variational evolution*.

Local minimality directly refers to a topology, whereas global minimality is topology-independent. But, even if the latter is called upon, the failure to impart upon test cracks a decent topology would dim the mathematical hope for an existence result. A natural candidate is the Hausdorff metric, defined for two closed sets A, B as

$$
d_H(A, B) := \max\{\sup_{a \in A} d(a, B), \sup_{b \in B} d(b, A)\}.
$$

Examine for instance the initial time in the global minimality context with $\Gamma_0 = \emptyset$, $f_b(0) = f_s(0) = 0$. Then, we should minimize

$$
\int_{\Omega \backslash \Gamma} W(\nabla \varphi) \, dx + k \mathcal{H}^1(\Gamma)
$$

among all pairs (Γ, φ) with $\varphi = g(0)$ on $\partial_d \Omega \backslash \Gamma$. The direct method of the calculus of variations would have us take an infimizing sequence $\{(\Gamma_n, \varphi_n)\}$. In particular, we are at liberty to assume that $\mathcal{H}^1(\Gamma_n) \leq C$. Say that the sequence Γ_n converges in the Hausdorff metric to some Γ; this is not a restriction, thanks to Blaschke's compactness theorem (Rogers, 1970). Then we would like to have

$$
\mathcal{H}^1(\Gamma) \leq \liminf_n \mathcal{H}^1(\Gamma_n).
$$

But, this is generically false, except in 2d and for, say, connected Γ_n's! Consequently, that topology seems a bit restrictive, although it has been used with success to prove existence, in the global minimality framework, for the 2d variational evolution restricted to connected cracks in (Dal Maso and Toader, 2002). We shall come back to this point in Section 5.

Light will shine from an unexpected direction. In the context of image segmentation, D. Mumford and J. Shah proposed to segment image through the following algorithm: Find a pair K, compact of $\Omega \subset \mathbb{R}^2$ (the picture) representing the contours of the image in the picture, and φ, the true pixel intensity at each point of the picture, an element of $C^1(\Omega \backslash K)$, which minimizes

$$\int_{\Omega \backslash K} |\nabla \varphi|^2 \, dx + k\mathcal{H}^1(K) + \int_\Omega |\varphi - g|^2 \, dx, \qquad (2.22)$$

where g is the measured pixel intensity. The minimization proposed in (Mumford and Shah, 1989) was then shown in (De Giorgi et al., 1989) to be equivalent to a well-posed one-field minimization problem on a subspace $SBV(\Omega)$ of the space $BV(\Omega)$ of functions with bounded variations on Ω, namely,

$$\int_\Omega |\nabla \varphi|^2 \, dx + k\mathcal{H}^1(S(\varphi)) + \int_\Omega |\varphi - g|^2 \, dx, \qquad (2.23)$$

where $\nabla \varphi$ represents the absolutely continuous part of the weak derivative of φ (a measure), and $S(\varphi)$ the set of jump points for φ.

We recall that a function $\varphi : \Omega \mapsto \mathbb{R}$ is in $BV(\Omega)$ iff $\varphi \in L^1(\Omega)$ and its distributional derivative $D\varphi$ is a measure with bounded total variation. Then, the theory developed by E. De Giorgi (see e.g. (Evans and Gariepy, 1992)) implies that

$$D\varphi = \nabla \varphi(x) \, dx + (\varphi^+(x) - \varphi^-(x))\nu(x)\mathcal{H}^1 \lfloor S(\varphi) + C(\varphi),$$

with $\nabla \varphi$, the approximate gradient, $\in L^1(\Omega)$ ($\nabla \varphi$ is no longer a gradient), $S(\varphi)$ the complement of the set of Lebesgue points of φ, a $\mathcal{H}^1$ σ–finite and countably 1-rectifiable set (a countable union of compacts included in C^1–hypersurfaces, up to a set of 0 $\mathcal{H}^1$–measure), $\nu(x)$ the common normal to all those hypersurfaces at a point $x \in S(\varphi)$, $\varphi^\pm(x)$ the values of $\varphi(x)$ "above and below" $S(\varphi)$, and $C(\varphi)$ a measure (the Cantor part) which is mutually singular with dx and with $\mathcal{H}^1$ (it only sees sets that have 0 Lebesgue–measure and infinite $\mathcal{H}^1$–measure). The subspace $SBV(\Omega)$ is that of those $\varphi \in BV(\Omega)$ such that $C(\varphi) \equiv 0$. It enjoys good compactness properties established in (Ambrosio, 1990), namely

$$\varphi_n \in SBV(\Omega) \text{ with } \begin{cases} \varphi_n \text{ bounded in } L^\infty(\Omega) \\ \nabla\varphi_n \text{ bounded in } L^q(\Omega;\mathbb{R}^2), \ q > 1 \\ \mathcal{H}^1(S(\varphi_n)) \text{ bounded in } \mathbb{R} \end{cases}$$

$$\exists\{\varphi_{k(n)}\} \subset \{\varphi_n\}, \exists\varphi \in SBV(\Omega) \text{ s.t.}$$
$$\begin{cases} \varphi_{k(n)} \to \varphi, \text{ strongly in } L^p(\Omega), \ p < \infty \\ \nabla\varphi_{k(n)} \rightharpoonup \nabla\varphi, \text{ weakly in } L^q(\Omega;\mathbb{R}^2) \\ \mathcal{H}^1(S(\varphi)) \le \liminf_n \mathcal{H}^1(S(\varphi_{k(n)})) \end{cases} \tag{2.24}$$

Thanks to Ambrosio's compactness result above, a simple argument of the direct method applied to (2.23) establishes existence of a minimizer φ_g for that functional. The further result that the pair $\left(\varphi_g, \overline{(S(\varphi_g))}\right)$ is a minimizer for (2.22) is highly non-trivial and makes up the bulk of (De Giorgi et al., 1989).

In De Giorgi's footstep, we thus reformulate the variational evolution in the weak functional framework of SBV, or rather of those functions that have all their components in SBV, the jump set $S(\varphi)$ becoming the union of the jump set of each component of φ. To do this, it is more convenient to view the hard device $g(t)$ as living on all of $\mathbb{R}^2$ and to integrate by parts the boundary term involving $\dot{g}(t)$ in (Eb). So, after elementary integrations by parts, we propose to investigate

- **The weak variational evolution** : Find $(\varphi(t), \Gamma(t))$ satisfying

(Ic) $\Gamma(0) = \Gamma_0$;

(Ulm) $(\Gamma(t), \varphi(t))$ is a local minimizer (in a topology that remains to be specified) for

$$\mathcal{E}(t; \varphi, \Gamma) := \int_\Omega W(\nabla\varphi) \, dx - \mathcal{F}(t, \varphi) + k\mathcal{H}^1(\Gamma), \tag{2.25}$$

among all $\overline{\Omega}\backslash\partial_s\Omega \supset \Gamma \supset \Gamma(t)$ and all $\varphi \equiv g(t)$ on $\mathbb{R}^2\backslash\overline{\Omega}$ with $S(\varphi) \subset \Gamma$; or, resp.,

(Ugm) $(\Gamma(t), \varphi(t))$ is a global minimizer for $\mathcal{E}(t; \varphi, \Gamma)$ among all $\overline{\Omega}\backslash\partial_s\Omega \supset \Gamma \supset \Gamma(t)$ and all $\varphi \equiv g(t)$ on $\mathbb{R}^2\backslash\overline{\Omega}$ with $S(\varphi) \subset \Gamma$;

(Eb) $\dfrac{dE}{dt}(t) = \displaystyle\int_\Omega \dfrac{\partial W}{\partial F}(\nabla\varphi(t)) \cdot \nabla\dot{g}(t) \, dx - \dot{\mathcal{F}}(t, \varphi(t)) - \mathcal{F}(t, \dot{g}(t))$

with

$$E(t) = \mathcal{E}(t; \varphi(t), \Gamma(t)).$$ (2.26)

The weak formulation calls for several caveats. The attentive reader will have remarked that, in spite of the previous considerations on SBV, we have not explicitly indicated where φ (or $\varphi(t)$) should live. This is because, when dealing with vector-valued SBV-functions (the case of plane (hyper)elasticity, for example), that space – that is the Cartesian product of SBV for each of the components – is not quite sufficient. One should really work in $GSBV$ (Dal Maso et al., 2005). Our narration of the variational evolution is mathematically precise, but an overload of technicalities would serve no useful purpose and the curious reader may wish to consult (Dal Maso et al., 2005) on this issue.

Likewise, it is not so that the crack should belong to $\overline{\Omega} \backslash \partial_s \Omega$. Any rigorous analysis will actually require $\partial_s \Omega$, the site of application of the surface forces, to be part of the boundary of a non-brittle piece of the material. In other words, we should in truth single out a thin layer around $\partial_s \Omega$ with infinite fracture toughness. This also will be overlooked in the sequel.

Also, for the mathematically-minded reader, the test cracks Γ do not have to be even essentially, *i.e.*, up to a set of $\mathcal{H}^1$-measure 0, closed subsets of $\overline{\Omega} \backslash \partial_s \Omega$, but only countably 1-rectifiable curves. Whether the actual crack $\Gamma(t)$ that could be produced through the weak variational evolution is closed or not will be deemed a question of regularity and briefly commented upon in Paragraph 5.1.4 in the setting of global minimization.

Then, observe that, when dealing with plane (hyper)elasticity, the surface energy does not force non-interpenetration of the crack lips, but counts all jumps, whether interpenetrating or not. As mentioned at the onset of this section, non-interpenetration is a delicate issue in Griffith's setting. We will stay clear of it for the remainder of these notes and be forced to accept the potential occurrence of interpenetration.

Finally, as before, the same labels have been kept. The context will clearly indicate if the relevant formulation is weak or strong.

REMARK 2.6. In practice, the reader should feel entitled to identify the crack $\Gamma(t)$ with $\Gamma_0 \cup \bigcup_{s \leq t} S(\varphi(s))$, although caution should be exercised in finite elasticity, in which case it is unclear whether that identification remains legitimate.

A few transgressions notwithstanding, the recasting of Griffith's evolution model in a variational framework is complete. Its success or failure hinges on its ability to perform when confronted with initiation, irreversibility and path. It will fail, more often than not, although the failure becomes a resounding success when gauged by the standards of the classical theories. The villain is easily identified: Griffith, or rather the form of the surface energy that was proposed by Griffith. The shortcomings of that energy have long been acknowledged and the idea of a cohesive type surface energy has since emerged, most notably in (Barenblatt, 1962), (Needleman, 1992).

In the next subsection, we examine how to import a Barenblatt type energy into the variational evolution.

2.6. COHESIVENESS AND THE VARIATIONAL EVOLUTION

Early on, it was recognized that inter-atomic bonds of the underlying lattice of an crystalline solid will "stretch" before they break, and thus that some degree of reversibility near the crack tip should precede the advance of a crack. In other words there is a barrier to bond break and that barrier can be thought of as a macroscopic manifestation of the elasticity of the underlying inter-atomic potential. In any case, such considerations have prompted the replacement of Griffith's surface energy by various surface energies that all share common defining features. They often read as

$$\int_{\text{crack}} \kappa \left(\left| [\varphi(s)] \right| \right) ds,$$

where $[\varphi(s)]$ stands for the jump of the field φ at the point with curvilinear abscissa s on the crack and κ is as in Figure 2.3.

Note the main ingredients. A concave increasing function which takes the value 0 at 0 and asymptotically converges to the value k of the fracture toughness. The slope at 0, σ_c is positive and finite.

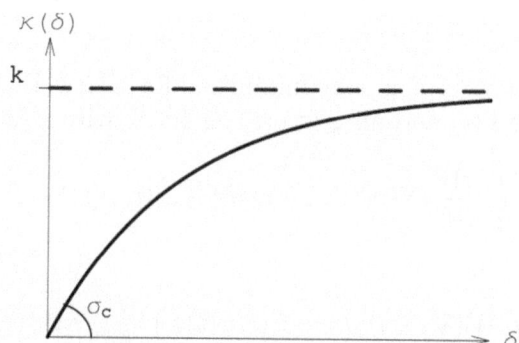

Figure 2.3. A typical cohesive surface energy

Here, non-interpenetration is not addressed in the vector-valued case since the Euclidean norm of the jump enters the expression for the surface energy. As previously announced, the issue will be tackled in Subsections 4.2, 5.2.

In the context of the weak variational evolution, we suggest to replace the term $k\mathcal{H}^1(\Gamma)$ in (2.25) by $\int_\Gamma \kappa(|[\varphi] \vee \psi(t)|)d\mathcal{H}^1$, where $\psi(t)$ is the cumulated jump, up to time t, at the given point of Γ. Otherwise said,

$$\psi(t) = \vee_{\tau \leq t}[\varphi(\tau)].$$

Also, the term $k\mathcal{H}^1(\Gamma)$ in (2.26) is replaced by $\int_{\Gamma(t)} \kappa(|\psi(t)|) d\mathcal{H}^1$. Let us explain our reasons. We postulate that the energy dissipated by the creation of discontinuities is only dissipated once for a given value of the jump, and that additional dissipation will only occur for greater mismatches between the lips of the incipient part of the crack. This is of course one of many plausible phenomenological assumptions; it all depends on what irreversibility means in a cohesive context! We will return to this issue in Section 5.

In any case, the ensuing formulation reads as follows:

- **The weak cohesive variational evolution** : Find, for every $t \in [0, T]$, $(\varphi(t), \Gamma(t))$ satisfying, with

$$\psi(t) := \vee_{\tau \leq t}[\varphi(\tau)], \tag{2.27}$$

(Ic) $\Gamma(0) = \Gamma_0$;

(Ulm) $(\Gamma(t), \varphi(t))$ is a local minimizer (in a topology that remains to be specified) for

$$\mathcal{E}(t; \varphi, \Gamma) := \int_\Omega W(\nabla\varphi)dx - \mathcal{F}(t, \varphi) + \int_\Gamma \kappa\left(\left|[\varphi] \vee \psi(t)\right|\right)d\mathcal{H}^1 \tag{2.28}$$

among all $\overline{\Omega} \backslash \partial_s \Omega \supset \Gamma \supset \Gamma(t)$ and all $\varphi \equiv g(t)$ on $\mathbb{R}^2 \backslash \overline{\Omega}$ with $S(\varphi) \subset \Gamma$; or, resp.,

(Ugm) $(\Gamma(t), \varphi(t))$ is a global minimizer for $\mathcal{E}(t; \varphi, \Gamma)$ among all $\overline{\Omega} \backslash \partial_s \Omega \supset \Gamma \supset \Gamma(t)$ and all $\varphi \equiv g(t)$ on $\mathbb{R}^2 \backslash \overline{\Omega}$ with $S(\varphi) \subset \Gamma$;

(Eb) $\dfrac{dE}{dt}(t) = \displaystyle\int_\Omega \dfrac{\partial W}{\partial F}(\nabla\varphi(t)).\nabla\dot{g}(t)\,dx - \dot{\mathcal{F}}(t, \varphi(t)) - \mathcal{F}(t, \dot{g}(t))$

with

$$E(t) = \int_\Omega W(\nabla\varphi(t))\,dx - \mathcal{F}(t, \varphi(t)) + \int_{\Gamma(t)} \kappa(|\psi(t)|)d\mathcal{H}^1. \tag{2.29}$$

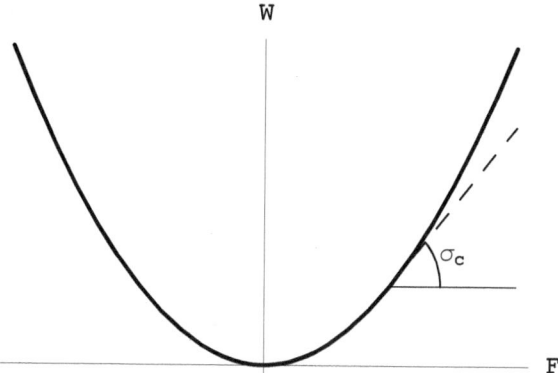

Figure 2.4. Original bulk energy density

The weak cohesive variational evolution, although formally resembling the weak variational formulation obtained in Subsection 2.5 (take $\kappa(0) = 0$ and $\kappa \equiv 1, \delta \neq 0$), yet it is mathematically more troublesome. Examine once again the initial time in the global minimality context with $\Gamma_0 = \emptyset$, $f_b(0) = 0$ and $\partial_d\Omega = \partial\Omega$, and also assume, for simplicity that $W(F) = 1/2|F|^2$ (see Figure 2.4).

Then, in the case of anti-plane shear, one should minimize

$$1/2 \int_\Omega |\nabla\varphi|^2 \, dx + \int_{S(\varphi)} \kappa\left(\left|[\varphi]\right|\right) d\mathcal{H}^1$$

among all φ's with $\varphi = g(0)$ on $\mathbb{R}^2 \backslash \overline{\Omega}$ (it is enough here to take $\Gamma = S(\varphi)$). In contrast to what was encountered when using Griffith's energy, the above functional *does not* admit a minimizer in $SBV(\mathbb{R}^2)$. A relaxation process occurs whereby, for high enough gradients, it is energetically cheaper to replace those by many infinitesimally small jumps; see (Bouchitté et al., 1995), (Braides et al., 1999), (Bouchitté et al., 2002). We will return to this point in Subsection 4.2 below. For now, we merely observe that the relaxed functional – that minimized at the limits of minimizing sequences for the original functional – will be of the form

$$\int_\Omega \hat{W}(\nabla\varphi) \, dx + \int_{S(\varphi)} \kappa\left(\left|[\varphi]\right|\right) d\mathcal{H}^1 + \sigma_c|C(\varphi)| \qquad (2.30)$$

with (see Figure 2.5)

$$\hat{W}(F) = \begin{cases} 1/2|F|^2, \text{ if } |F| \leq \sigma_c \\ 1/2(\sigma_c)^2 + \sigma_c(|F| - \sigma_c) \text{ otherwise,} \end{cases}$$

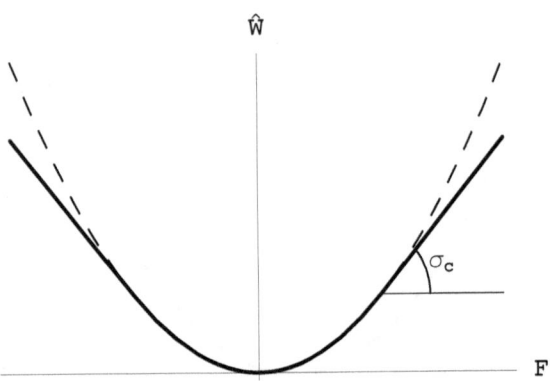

Figure 2.5. Relaxed bulk energy density

and where $|C(\varphi)|$ stands for the total variation of the measure $C(\varphi)$ (see e.g. (Ambrosio et al., 2000), Section 5.5). The resulting functional has *linear* growth at infinity and it does not live on SBV but on BV (because of the reappearance of a Cantor part)!

The output of the minimization is no longer a crack, because the Cantor part corresponds in effect to some kind of "diffuse cracking process" with overall dimensionality higher than 1 in 2d. It is not so clear how one should proceed onward, unless minima of the relaxed functional are actually in SBV. This is true in 1d, as demonstrated in (Braides et al., 1999), but only wishful thinking in higher dimensions.

As we see, the introduction of cohesive surface energies enriches the model, but it also bears its share of misfortunes. The subsequent developments will attempt in part to weigh the respective merits of both Griffith and cohesive approaches within the adopted framework.

3. Stationarity versus local or global minimality – A comparison

Before embarking on the variational journey, we wish to explore the ramifications of minimality in the context of both Griffith and cohesive fracture. The adopted setting, or rather settings, for such an analysis are designed so that the "crack path" is not at stake. Nor is irreversibility a concern here because the monotonicity of the loads combined with the geometry of the problems result in an increase of both the measure of the discontinuity set and the magnitude of the discontinuities on that set. The focus is squarely on minimality, although, at times energy balance (Eb) will come to the rescue.

The two settings are

1. A 1d-traction experiment under a hard or a soft device;

2. A 2d-tearing experiment.

In the first setting, cracks are merely points of discontinuity along the bar; in the second setting, symmetry of the geometry and of the loads suggests a straight crack path in mode III. In both settings, we assess the potential existence of weak variational evolutions satisfying unilateral stationarity (Ust), unilateral minimality (Ulm), or still unilateral global minimality (Ugm), together with energy balance (Eb), this for both Griffith, or cohesive fracture energies. The resulting picture is a dizzying labyrinth, but maybe it is because we have "realized that [fracture] and the labyrinth were one and the same"[6].

3.1. 1D TRACTION

A "crack-free" homogeneous linearly elastic bar of length L, cross-sectional area Σ, Young's modulus E, toughness k is clamped at $x = 0$ and subject to a displacement load εL, $\varepsilon \nearrow$ (hard device), or to a force load $\sigma \Sigma$, $\sigma \nearrow$ (soft device) at $x = L$. The parameters σ, ε play the role of the time variable. Thus, all evolutions will be parameterized by either σ, or ε.

The results are concatenated in Conclusions 3.2, 3.3, 3.4, 3.6 and do support the labyrinthine paradigm. Those of the cohesive case were first partially obtained in (Del Piero, 1997) and also analyzed in (Braides et al., 1999).

3.1.1. *The Griffith case – Soft device*

Assume that φ is an admissible deformation field for a value σ of the loading parameter; that field may have jumps $S(\varphi) \subset [0, L)$, or it may

---

[6] adapted from: Jorge Luis Borges – The Garden of Forking Paths

correspond to the elastic state, in which case it lies in $W^{1,2}(0,L)$. In any case we view it as a field defined in $SBV(\mathbb{R})$ and such that $\varphi \equiv 0$ on $(-\infty, 0)$. Its associated energy is

$$\mathcal{E}(\sigma, \varphi) = \frac{1}{2} \int_{(0,L)} E\Sigma(\varphi' - 1)^2 \, dx - \sigma\Sigma\varphi(L+) + k\Sigma\#(S(\varphi)), \quad (3.1)$$

and that energy will only be finite if $S(\varphi)$ is finite and $\varphi' \in L^2(0,L)$, which we assume from now onward. This in turn implies that we may as well restrict the admissible fields to be in $SBV(\mathbb{R}) \cap L^\infty(\mathbb{R})$.

Consider the perturbation $\varphi + h\zeta$ with ζ admissible. The corresponding energy is

$$\mathcal{E}(\sigma, \varphi + h\zeta) = \mathcal{E}(\sigma, \varphi) + k\Sigma\#(S(\zeta)\backslash S(\varphi)) - h\sigma\Sigma\zeta(L+) \quad (3.2)$$

$$+ h\int_{(0,L)} E\Sigma(\varphi' - 1)\zeta' \, dx + \frac{h^2}{2}\int_{(0,L)} E\Sigma\zeta'^2 \, dx.$$

Then $\mathcal{E}(\sigma, \varphi + h\zeta) > \mathcal{E}(\sigma, \varphi)$ as soon as $S(\zeta)\backslash S(\varphi) \neq \emptyset$ and h is small enough. Thus unilateral stationarity need only be checked when $S(\zeta) \subset S(\varphi)$.

Then, (Ust) yields

$$0 \leq \int_{(0,L)} E\Sigma(\varphi' - 1)\zeta' \, dx - \sigma\Sigma\zeta(L+),$$

for all admissible ζ's, or still

$$E(\varphi' - 1) = \sigma \quad \text{in } (0,L), \quad E(\varphi' - 1) = 0 \quad \text{on } S(\varphi).$$

Whenever $\sigma > 0$, only the elastic deformation $\varphi_e(\sigma)(x) = x + \sigma x/E$ satisfies unilateral stationarity. In turn, (3.2) yields

$$\mathcal{E}(\sigma, \varphi_e(\sigma) + h\zeta) - \mathcal{E}(\sigma, \varphi_e(\sigma)) = \frac{h^2}{2}\int_0^L E\Sigma\zeta'^2 \, dx$$

$$+ \Sigma\left\{ k\#(S(\zeta)) - h\sigma\sum_{S(\zeta)}[\zeta] \right\} \geq 0,$$

provided that h is small compared to any norm that controls $\sum_{S(\zeta)}[\zeta]$ (the sup-norm or the BV-norm for example). This ensures local minimality of the elastic solution in any topology associated with such norms.

The elastic solution cannot be a global minimum when $\sigma > 0$, because the energy given by (3.1) is not bounded from below: just take $\varphi(x) = x$ in $[0, L/2)$ and $\varphi(x) = x + n$ in $(L/2, L]$. Global minimality behaves very erratically when confronted with soft devices. This major drawback will be further dissected in Paragraph 4.1.1.

REMARK 3.1. Testing the elastic solution against non-interpenetrating jumps is easy, since it suffices to restrict test jumps to be non-negative. In this context, the elastic solution is checked to be a global minimum for $\sigma < 0$, if non-interpenetration is imposed.

In conclusion, under a soft device, the elastic configuration is the only one that satisfies (Ust) and/or (Ulm). Because energy balance is automatic in the case of a purely elastic evolution, we thus conclude that

CONCLUSION 3.2. *In a 1d traction experiment with a soft device, the elastic evolution is the only one that satisfies the weak variational evolution with either* (Ust), *or* (Ulm), *and* (Eb). *There is no solution to the weak variational evolution with* (Ugm) *and* (Eb).

3.1.2. *The Griffith case – Hard device*

The admissible deformations are still in $SBV(\mathbb{R})$ and they satisfy $\varphi \equiv 0$ on $(-\infty, 0)$ and $\varphi \equiv (1+\varepsilon)L$ on (L, ∞). The associated energy is

$$\mathcal{E}(\varepsilon, \varphi) = \frac{1}{2} \int_{(0,L)} E\Sigma(\varphi' - 1)^2 \, dx + k\Sigma\#(S(\varphi)), \qquad (3.3)$$

and, once again it is only finite if $\#(S(\varphi))$ is finite and $\varphi' \in L^2(0, L)$, which we assume.

The argument is very close to that of the previous paragraph and uses similar test functions. Unilateral stationarity (Ust) yields

$$E(\varphi' - 1) = \sigma \quad \text{in } (0, L), \quad E(\varphi' - 1) = 0 \quad \text{on } S(\varphi),$$

where σ is now an unknown constant.

If $\sigma \neq 0$, then $S(\varphi) = \emptyset$ and $\varphi \equiv \varphi_e(\varepsilon)$, the elastic response; $\varphi_e(\varepsilon)(x) = (1+\varepsilon)x$ and $\sigma = E\varepsilon$. The associated energy is $\mathcal{E}(\varphi_e(\varepsilon)) = E\Sigma L\varepsilon^2/2$ and, as in the case of a soft device, it is a local minimum for similar topologies.

If now $\sigma = 0$, then $S(\varphi) \neq \emptyset$, otherwise $\varphi \equiv \varphi_e(\varepsilon)$. For a given number j of jumps, the field $\varphi_j(\varepsilon)$ must be such that

$$(\varphi_j(\varepsilon))' = 1 \text{ in } [0, L] \backslash S(\varphi_j(\varepsilon)); \quad \sum_{S(\varphi_j(\varepsilon))} [\varphi_j(\varepsilon)] = \varepsilon L,$$

and the associated energy is $\mathcal{E}(\varphi_j(\varepsilon)) = kj\Sigma$.

Further, for any admissible ζ,

$$\mathcal{E}(\varepsilon, \varphi_j(\varepsilon) + h\zeta) - \mathcal{E}(\varepsilon, \varphi_j(\varepsilon)) = \frac{h^2}{2} \int_{(0,L)} E\Sigma\zeta'^2 \, dx \geq 0,$$

which ensures that $\varphi_j(\varepsilon)$ is a local minimum for similar topologies.

This time, the energy is bounded from below and a global minimum $\varphi_g(\varepsilon)$ exists. An immediate computation shows that

$$\varphi_g(\varepsilon) = \begin{cases} \varphi_e(\varepsilon) \text{ if } 0 < \varepsilon \leq \sqrt{2k/EL} \\ \varphi_1(\varepsilon) \text{ if } \varepsilon \geq \sqrt{2k/EL}. \end{cases} \tag{3.4}$$

As far as the energy balance (Eb) is concerned, the elastic solution satisfies it automatically. Since the energy associated with any of the fields $\varphi_j(\varepsilon)$ is constant, while the associated stress is 0 throughout the bar, (Eb) is also satisfied by $\varphi_j(\varepsilon)$, since there are no contributions of (2.3),(2.9). Finally φ_g^e satisfies (Eb) for similar reasons.

CONCLUSION 3.3. *In a 1d traction experiment with a hard device, the elastic evolution, and all admissible evolutions with a set finite number of jumps satisfy the weak variational evolution with* (Ulm) *– and also* (Ust) *– and* (Eb). *Only $\varphi_g(\varepsilon)$ defined in (3.4) satisfies the weak variational evolution with* (Ugm) *and* (Eb).

Also, all evolutions that are elastic, up to $\varepsilon = \sqrt{2ik/EL}$, then have i jumps satisfy (Ulm) *– and also* (Ust) *– and* (Eb).

3.1.3. *Cohesive case – Soft device*

The surface energy κ has to be specified. We assume that

$$\begin{cases} \kappa \in \mathcal{C}^\infty, \text{ is strictly monotonically increasing, strictly concave on } \mathbb{R}^+ \\ (\kappa')^{-1} \text{ is convex} \\ \kappa(0) = 0, \ \kappa(\infty) = k, \ \kappa''(\infty) = 0, \ \kappa'(0) := \sigma_c > 0. \end{cases}$$
$$\tag{3.5}$$

As will be further dwelt upon in Subsection 4.2, non-interpenetration is readily imposed in the cohesive setting, so that, in analogy with Paragraph 3.1.2, the admissible fields φ will be elements of $SBV(\mathbb{R}) \cap L^\infty(\mathbb{R})$ such that $\varphi \equiv 0$ on $(-\infty, 0)$, $S(\varphi) \subset [0, L]$, $[\varphi] \geq 0$ on $S(\varphi)$. The energy reads as

$$\mathcal{E}(\sigma, \varphi) = \frac{1}{2} \int_{(0,L)} E\Sigma(\varphi' - 1)^2 \, dx - \sigma\Sigma\varphi(L) + \sum_{S(\varphi)} \kappa([\varphi])\Sigma. \tag{3.6}$$

Take an admissible test field $\zeta \in SBV(\mathbb{R}) \cap L^\infty(\mathbb{R})$; it satisfies $\zeta \equiv 0$ on $(-\infty, 0)$, $S(\zeta) \subset [0, L)$, $[\zeta] > 0$ on $S(\zeta) \backslash S(\varphi)$. Unilateral stationarity (Ust) then is easily seen to be equivalent to

$$\int_{(0,L)} E\Sigma(\varphi' - 1)\zeta' \, dx - \sigma\Sigma\zeta(L+) + \Sigma \sum_{S(\zeta) \cap S(\varphi)} \kappa'([\varphi])[\zeta] + \sigma_c\Sigma \sum_{S(\zeta)\backslash S(\varphi)} [\zeta] \geq 0.$$
$$\tag{3.7}$$

This is in turn equivalent to

$$
\begin{aligned}
E(\varphi' - 1) &= \sigma \quad \text{in } (0, L), \\
\kappa'([\varphi]) &= \sigma \quad \text{on } S(\varphi) \\
\sigma &\leq \sigma_c.
\end{aligned}
\tag{3.8}
$$

We borrow the derivation of (3.8) from (Braides et al., 1999), Section 6. First, assume that $E(\varphi' - 1) \neq cst.$ on $(0, L)$. Then, there exist $c < d$ such that $A_c := \{E(\varphi' - 1) \leq c\}$ and $A_d := \{E(\varphi' - 1) \geq d\}$ have positive measure. Take $\zeta := \int_0^x \{|A_d|\chi_{A_c} - |A_c|\chi_{A_d}\} (s)ds$, so that, replacing in (3.7), we get

$$
|A_c||A_d|(c - d) \geq 0,
$$

a contradiction, hence the first condition.

Now, take $\zeta(x) := -\dfrac{x}{L}$, $x < x_0$ and $\zeta(x) := 1 - \dfrac{x}{L}$, $x > x_0$ with $x_0 \in [0, L]$. From (3.7), we get, in view of the already established first condition in (3.8),

$$
\sigma \leq \kappa'([\varphi(x_0)]),
$$

whenever $x_0 \in S(\varphi)$. But then, $-\zeta$ is also an admissible test, so that we get the opposite inequality, hence the second condition in (3.8). Consequently, as soon as $\sigma \neq 0$, there can only be a finite number of jump points. The third condition is obtained similarly, taking a point $x_0 \notin S(\varphi)$.

For $0 < \sigma < \sigma_c$, the above conditions are met by infinitely many configurations, the elastic solution $\varphi_e(\sigma)$ for one, then any solution

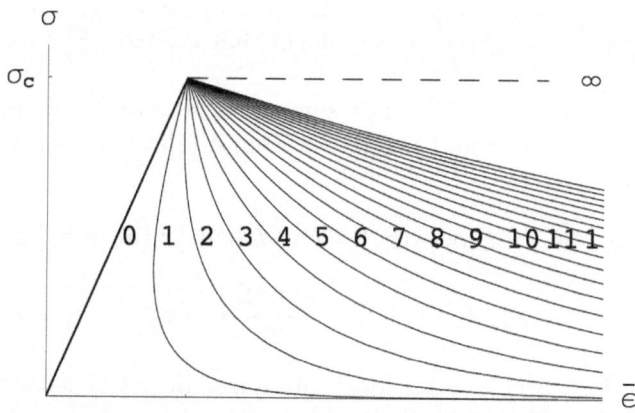

Figure 3.1. 1d traction – stationary solutions – i denotes the number of discontinuity points

$\varphi_i(\sigma)$ with

$$
\begin{cases}
(\varphi_i(\sigma))' = 1 + \dfrac{\sigma}{E} \\
\#(S(\varphi_i(\sigma))) = i \\
[\varphi_i(\sigma)](x) = (\kappa')^{-1}(\sigma), \ x \in S(\varphi_i(\sigma)).
\end{cases}
\tag{3.9}
$$

The average deformation $\bar{\epsilon}$ of the bar - in mathematical terms, the total variation of $(\varphi_i(\sigma) - x)$ – is given by

$$
\bar{\epsilon}L = \int_{(0,L)} ((\varphi_i(\sigma))' - 1)\, dx + \sum_{S(\varphi_i(\sigma))} [\varphi_i(\sigma)] = L\frac{\sigma}{E} + i(\kappa')^{-1}(\sigma).
$$

Hence,

$$
\bar{\epsilon} = \frac{\sigma}{E} + i\frac{(\kappa')^{-1}(\sigma)}{L},
\tag{3.10}
$$

which represents a one-parameter family of curves indexed by i, see Figure 3.1.

The elastic evolution satisfies (Ulm) for many reasonable topologies. Indeed, since, for any admissible test field ζ,

$$
\zeta(L) - \zeta(0^-) = \int_{(0.L)} \zeta'\, dx + \sum_{S(\zeta)} [\zeta],
$$

a straightforward computation leads to

$$
\mathcal{E}(\sigma, \varphi_e(\sigma)+h\zeta) = \mathcal{E}(\sigma, \varphi_e(\sigma)) + \sum_{S(\zeta)} (\kappa(h[\zeta]) - h\sigma[\zeta])\Sigma + \frac{h^2}{2}\int_0^L E\Sigma\zeta'^2\, dx.
$$

Since $\zeta \in L^\infty(\mathbb{R})$, we conclude that $\mathcal{E}(\sigma, \varphi_e(\sigma) + h\zeta) \geq \mathcal{E}(\sigma, \varphi_e(\sigma))$ for h small enough compared to any norm that controls $\sum_{S(\zeta)} [\zeta]$, as long as $\sigma < \sigma_c$.

Take an evolution $\varphi_i(\sigma)$. For some $x \in S(\varphi_i(\sigma))$, choose the test field ζ with $\zeta' = 0$ in $(0, L)$, $\zeta(0-) = 0$ and $[\zeta](x) = -L$. Then, for h small enough,

$$
\mathcal{E}(\sigma, \varphi_i(\sigma) + h\zeta) - \mathcal{E}(\sigma, \varphi_i(\sigma)) = h\sigma\Sigma L + \kappa([\varphi_i(\sigma) + h\zeta](x))\Sigma
$$

$$
-\kappa([\varphi_i(\sigma)](x))\Sigma = \frac{h^2 L^2 \Sigma}{2}\kappa''((\kappa')^{-1}(\sigma)) + o(h^2) < 0.
$$

Consequently, $\varphi_i(\sigma)$ is not a local minimizer in any reasonable norm.

The term $-\sigma\Sigma\varphi(L)$ in (3.6), together with the upper bound k on the surface energy prevent the total energy from being bounded from below; no evolution satisfies (Ugm).

As always, the elastic evolution satisfies (Eb). The total energy associated with $\varphi_i(\sigma)$ is given by

$$\mathcal{E}(\sigma, \varphi_i(\sigma)) = -\frac{1}{2}\frac{\sigma^2 \Sigma L}{E} - \Sigma \sigma L + i\Sigma \left\{ \kappa\left((\kappa')^{-1}(\sigma)\right) - \sigma(\kappa')^{-1}(\sigma) \right\},$$

while the only non-zero term in the right hand side of (Eb), $\dot{\mathcal{F}}(\sigma, \varphi_i(\sigma))$ (see (2.9)) reduces to

$$\Sigma L(1 + \bar{\varepsilon}) = \Sigma \left\{ L + \frac{\sigma L}{E} + i(\kappa')^{-1}(\sigma) \right\}.$$

From this, equality in the balance of energy becomes immediate by e.g. derivation of $\mathcal{E}(\sigma, \varphi_i(\sigma))$ with respect to σ.

CONCLUSION 3.4. *In a 1d traction experiment with a soft device and as long as $\sigma < \sigma_c$, the elastic evolution is the only one that satisfies the weak variational evolution with (Ulm) and (Eb). During that time interval, all evolutions given by (3.9) satisfy the weak variational evolution with (Ust) and (Eb). There are no evolutions satisfying (Ugm) and (Eb).*

It is not possible in the present context to jump from the elastic solution to one of the evolutions given by (3.9), or from one of those to one with a different number of jumps because the total energy does not remain continuous at such a jump, but increases brutally through that jump; thus (Eb) is not satisfied.

REMARK 3.5.

a. Note that the evolution branches corresponding to (3.9) for $j \neq 0$ start with infinite average deformation at $\sigma = 0$! But this is not admissible from the standpoint of the ambient space. Since our self-imposed rule is to start within a space of functions with bounded variations, we thus have to reject those solutions, unless we agree that, at initial time, $\sigma > 0$. Further, it is not possible to jump onto one of those branches without contravening energy balance. So, the alternative in the case of a cohesive evolution with a soft device is clear. If starting from an unloaded configuration, the elastic evolution is the only one that respects (Eb) for $\sigma < \sigma_c$. Transgression is punished by an increase in total energy;

b. The ever attentive reader will wonder, with some cause, what happens when $\sigma \geq \sigma_c$. We share her interrogations and merely refer her to Remark 4.10 as a ground for possible future investigations of the post-critical case.

3.1.4. Cohesive case – Hard device

The surface energy is again described by (3.5). The admissible fields φ will be elements of $SBV(\mathbb{R}) \cap L^\infty(\mathbb{R})$ such that $\varphi \equiv 0$ on $(-\infty, 0)$, $\varphi \equiv (1+\varepsilon)L$ on (L, ∞), $S(\varphi) \subset [0, L]$, $[\varphi] \geq 0$ on $S(\varphi)$. The associated energy is

$$\mathcal{E}(\varepsilon, \varphi) = \frac{1}{2} \int_{(0,L)} E\Sigma(\varphi' - 1)^2 \, dx + \sum_{S(\varphi)} \kappa([\varphi])\Sigma. \qquad (3.11)$$

As in the case of a soft device treated in Paragraph 3.1.3, unilateral stationarity (Ust) is equivalent to (3.8), although σ, which is still a constant, is not a datum anymore. Since ε is the average deformation, that is $\bar{\varepsilon} = \varepsilon$, an evolution $\varphi_i(\varepsilon)$ with i jumps must be such that σ satisfies (3.10). The point $(\bar{\varepsilon} = \varepsilon, \sigma)$ is still on one of the curves in Figure 3.1. But, in such a case, the evolution cannot start with $\varepsilon = 0$, because it would not satisfy (3.10). It has to start at

$$\varepsilon \geq \varepsilon_{i0} := \min \{\varepsilon; (\varepsilon, \sigma) \text{ satisfies } (3.10)\}. \qquad (3.12)$$

That minimum exists because $\varepsilon(\sigma)$ given by (3.10) is convex, since, by assumption (3.5), $(\kappa')^{-1}$ is convex, $\varepsilon(\sigma = 0) = \infty$ and $\varepsilon(\sigma = \sigma_c) = \sigma_c/E =: \varepsilon_c$. Note that it may be the case that $\varepsilon_{i0} = \varepsilon_c$.

The elastic evolution $\varphi_e(\varepsilon)(x) = (1 + \varepsilon)x/L$, for which $\sigma = E\varepsilon$, satisfies (Ulm) for many reasonable topologies, as long as $\varepsilon < \sigma_c/E$. Indeed, for any admissible test field ζ,

$$0 = \int_{(0.L)} \zeta' dx + \sum_{S(\zeta)} [\zeta],$$

so that

$$\mathcal{E}(\varepsilon, \varphi_e(\varepsilon) + h\zeta) - \mathcal{E}(\sigma, \varphi_e(\varepsilon)) = \sum_{S(\zeta)} (\kappa(h[\zeta]) - hE\varepsilon[\zeta])\Sigma$$
$$+ \frac{h^2}{2} \int_{(0,L)} E\Sigma\zeta'^2 \, dx.$$

This last expression remains non-negative for h small enough, provided that $\varepsilon < \sigma_c/E$.

Now, any evolution with 2 discontinuity points or more cannot satisfy (Ulm). Indeed, take $x_1 \neq x_2$ to be in the jump set of the evolution $\varphi_i(\varepsilon)$ corresponding to $i \geq 2$ discontinuity points. Take ζ to be such that $\zeta' = 0$ in $(0, L)$, $\zeta(0-) = 0$ et $[\zeta](x_1) = -[\zeta](x_2) = -L/2$. Then, for h small enough,

$$\mathcal{E}(\varepsilon, \varphi_i(\varepsilon) + h\zeta) - \mathcal{E}(\varepsilon, \varphi_i(\varepsilon)) = \Sigma\Big(\sum_{i=1,2} \kappa([(\varphi_i(\varepsilon) + h\zeta)(x_i)])$$

$$-\kappa([\varphi_i(\varepsilon)(x_i)])\Big) = \frac{h^2 \Sigma L^2}{4} \sum_{i=1,2} \kappa''((\kappa')^{-1}(\sigma)) + o(h^2) < 0.$$

The local minimality of $\varphi_1(\varepsilon)$ with discontinuity point x_1 will be ensured, provided that, for h small enough,

$$\mathcal{E}(\varepsilon, \varphi_1(\varepsilon) + h\zeta) - \mathcal{E}(\varepsilon, \varphi_1(\varepsilon)) = h \sum_{S(\zeta)\backslash\{x_1\}} (\sigma_c - \sigma)\Sigma[\zeta] + \frac{h^2}{2}\int_{(0,L)} E\Sigma\zeta'^2 dx$$

$$+ \frac{h^2}{2}\kappa''((\kappa')^{-1}(\sigma))\Sigma[\zeta(x_1)]^2 + o(h^2) \geq 0.$$

If $\sigma < \sigma_c$, the only possible challenge to local minimality will be from the term $\int_{(0,L)} E\zeta'^2 dx + \kappa''((\kappa')^{-1}(\sigma))[\zeta(x_1)]^2$, which must remain non-negative for all admissible ζ's with $S(\zeta) = \{x_1\}$. Let us compute

$$\lambda = \min\Big\{\int_{(0,L)} E\zeta'^2 dx \mid \zeta \in SBV \cap L^\infty(\mathbb{R}); \zeta(0^-) = \zeta(L^+) = 0;$$

$$[\zeta(x_1)] = 1, \ S(\zeta) = \{x_1\}\Big\}.$$

It is easily checked that $\lambda = E/L$, so that $\varphi_1(\varepsilon)$ is a local minimizer, for $\sigma < \sigma_c$ if, and only if

$$\frac{E}{L} + \kappa''((\kappa')^{-1}(\sigma)) \geq 0.$$

The graphic interpretation is simple. In view of (3.10) specialized to $i = 1$,

$$L\frac{d\varepsilon}{d\sigma} = \frac{L}{E} + \frac{1}{\kappa''((\kappa')^{-1}(\sigma))}, \tag{3.13}$$

so, since $\kappa'' < 0$, local minimality is equivalent to $\dfrac{d\varepsilon}{d\sigma} \leq 0$, or still, in view of (3.12) to $\varepsilon \geq \varepsilon_{10}$. For example, points close to $\sigma = \sigma_c$ are local minima if $L < -E/\kappa''(0)$ and are not local minima if $L > -E/\kappa''(0)$; observe that the minimality condition is length-dependent!

In any case the local minima correspond to evolutions that stay on the continuous lines in Figure 3.2.

Once again, the elastic evolution automatically satisfies (Eb). The total energy associated with $\varphi_i(\varepsilon)$, which only makes sense for $\varepsilon \geq \varepsilon_{i0}$

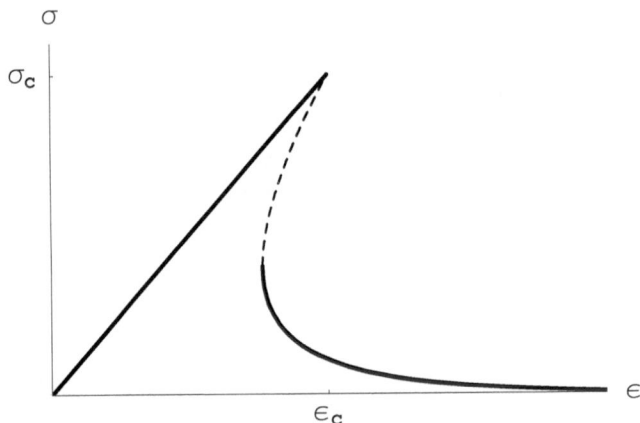

Figure 3.2. Locus of possible evolutions satisfying local minimality – at most 1 discontinuity point

(see (3.12)) is given by

$$\mathcal{E}(\varepsilon, \varphi_i(\varepsilon)) = \frac{1}{2}\frac{\sigma^2 \Sigma L}{E} + i\Sigma \left\{ \kappa \left((\kappa')^{-1}(\sigma) \right) \right\},$$

with σ related to ε by (3.10), while the only non-zero term in the right hand side of (**Eb**) is $\Sigma L \int_{\varepsilon_{i0}}^{\varepsilon} \sigma d\varepsilon$.

In view of (3.10), the derivatives in ε of the two quantities are seen to be equal. Equality in the balance of energy from ε_{i0} to ε is established. The fact that the relation between ε and σ in (3.10) can be inverted, and thus that σ is a well-defined function of ε has been implicitly used. This amounts to choosing a branch of the curve corresponding to i jumps in Figure 3.1 and to remain committed to it for the remainder of the evolution.

It is sometimes possible in the current context to jump from the elastic evolution to one of the evolutions $\varphi_i(\varepsilon)$. This will happen whenever there is a snap-back in the curve for $\varphi_i(\varepsilon)$ – that is whenever $\varepsilon_{i0} < \varepsilon_c$ – and the jump will occur at $\varepsilon = \varepsilon_i^* > \varepsilon_{i0}$. Indeed in such a case the total energy will remain continuous at such a jump, thus (**Eb**) is satisfied, being satisfied for each branch. To see this, just observe that the total energy corresponding to the evolution $\varphi_i(\varepsilon)$, in the case of a snap-back, corresponds to the area "under the curve", a quantity which is computed graphically as shown in Figure 3.3; see (Charlotte et al., 2000).

Then see Figure 3.4 for the determination of ε_i^* in the case where $\varepsilon_{i0} < \varepsilon_c$.

CONCLUSION 3.6. *In a 1d traction experiment with a hard device and as long as $\sigma < \sigma_c$, all evolutions given by (3.9), with σ related*

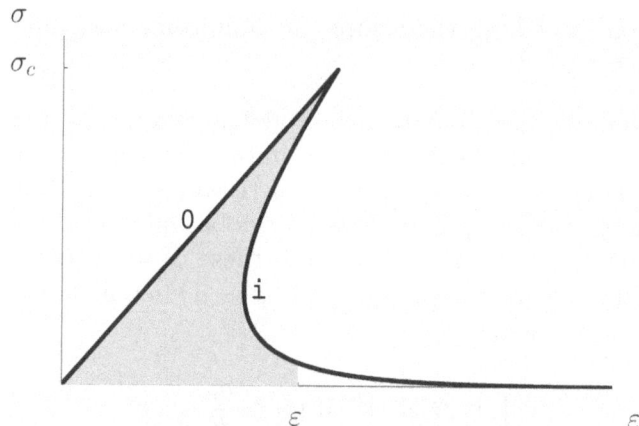

Figure 3.3. The total energy viewed as the area *under the curve*

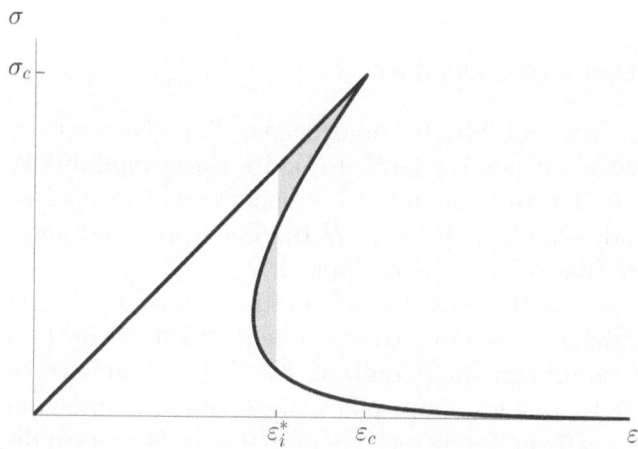

Figure 3.4. Determination of the jump deformation ε_i^* – the shaded surfaces have equal area

to ε by (3.10) satisfy the weak variational evolution with (**Ust**) and (**Eb**). The elastic evolution is the only one that always satisfies the weak variational evolution with (**Ulm**) and (**Eb**). The evolution $\varphi_1(\varepsilon)$ satisfies (**Ulm**) and (**Eb**) provided that $\dfrac{d\varepsilon}{d\sigma} \leq 0$, see Figure 3.2.

It is possible to jump from the elastic branch to a branch with i discontinuity points, provided that $\varepsilon_c > \varepsilon_{i0}$.

REMARK 3.7. The difference with the case of a soft device is striking because, here, we may start elastically with 0-deformation load, then jump onto an evolution with one, or more discontinuities, and this may be done even in the context of reasonable local minimality. Then we can keep on stretching the sample ad infinitum. Softening occurs as $\sigma \searrow 0$ when $\varepsilon \nearrow \infty$.

Note that we have not broached global minimality issues in this subsection. They are best left alone until Paragraph 4.2.3. However, it should be noted that it is proved in (Braides et al., 1999), Section 6, that the stationary points, local, or global minimizers of the original functionals $\mathcal{E}(\sigma$ resp. $\varepsilon, \varphi)$ are also those of its relaxation $\hat{\mathcal{E}}$ given by replacing the bulk term $\int_{(0,L)} E\Sigma(\varphi' - 1)^2 \, dx$ in (3.6), (3.11) by $\int_{(0,L)} \hat{W}(\varphi' - 1) \, dx$, with

$$\hat{W}(f) := \begin{cases} 1/2E\Sigma|f|^2, & \text{if } |f| \leq \dfrac{\sigma_c}{E} \\[2mm] 1/2\dfrac{(\sigma_c)^2\Sigma}{E} + \sigma_c\Sigma(|f| - \dfrac{\sigma_c}{E}) & \text{otherwise.} \end{cases}$$

3.2. A TEARING EXPERIMENT

Consider a thin semi-infinite homogeneous, linearly elastic slab of thickness $2H$, $\Omega = (0, +\infty) \times (-H, +H)$. Its shear modulus is μ and its toughness k. Tearing amounts to a displacement load $tH\mathbf{e}_3$ on $\{0\} \times (0, +H)$ and $-tH\mathbf{e}_3$ on $\{0\} \times (-H, 0)$. The upper and lower edges are traction free and no forces are applied.

We assume throughout that all solutions respect geometric symmetry, emphasizing that doing so cannot be justified; see in this respect the numerical experiment in Paragraph 8.3.2. The symmetry assumption permits one to look for a anti-plane shear solution, antisymmetric with respect to $y = 0$ and for a crack along that axis. We seek a displacement solution field of the form

$$\mathbf{u}(x, y, t) = \text{sign}(y)u(t, x)\mathbf{e}_3 \quad \text{with} \quad u(t, 0) = tH \tag{3.14}$$

and note that such a displacement cannot be the exact solution, because it fails to ensure the continuity of the normal stress at the points $(l(t), y)$, $y \neq 0$ (see (3.15)). The true symmetric solution can only be evaluated numerically, but it will be close to the proposed approximate solution as H becomes large or small.

The deformation $\varphi(t) = x + u(t)$ will be discontinuous at the points x on the $y = 0$-axis where $u(x, t) \neq 0$, that is

$$S(\varphi(t)) = \{x \geq 0 \ : \ \varphi(t, x) \neq x\}.$$

Then, the energy has the form

$$E(\varphi) = \int_0^\infty \mu H(\varphi'(x) - 1)^2 dx + \int_0^\infty \kappa\big(2\,|\varphi(x) - x|\big)dx,$$

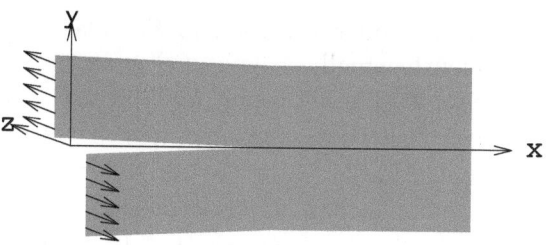

Figure 3.5. Tearing

where κ is the surface energy density. For Griffith's model, κ is discontinuous at zero and is k elsewhere, while the cohesive model calls for a differentiable, monotonically increasing κ with

$$\kappa(0) = 0, \quad \kappa(\delta) > 0 \text{ when } \delta > 0, \quad \kappa(\infty) = k.$$

In both settings, k represents the (tangential) toughness of the interface. For the Barenblatt model $\tau_c := \kappa'(0)$ is the ultimate shear stress, either finite ("initially rigid" cohesive response) or zero ("initially elastic" cohesive response).

The kinematically admissible test fields u ($\varphi = x + u$) at time t are elements of $W^{1,2}(0, +\infty)$ and satisfy $u(t, 0) = tH$. A global minimum for $\mathcal{E}$ exists for each t by elementary lower semi-continuity properties. We propose to show that (Ust), (Eb) has a unique solution, which identifies with the global minimum for $\mathcal{E}$ at t, which is thus unique.

Fix t. First, if $\varphi(t) = x + u(t)$ is solution to (Ust), then $u(t)$ is monotonically decreasing in x. Indeed, assume that a and b with $0 \leq a < b$ are such that $u(t, a) = u(t, b)$. Take v with $v = -u(t)$ in (a, b) and $v = 0$ otherwise. For $h \in (0, 1)$, $\varphi + hv$ is an admissible test function and

$$\mathcal{E}(\varphi + hv) - \mathcal{E}(\varphi) = (-2h + h^2) \int_a^b u'(t, x)^2 \, dx$$
$$+ \int_a^b \left(\kappa(2(1 - h) |u(t, x)|) - \kappa(2 |u(t, x)|) \right) dx$$
$$\leq (-2h + h^2) \int_a^b u'(t, x)^2 \, dx,$$

since κ is monotonically increasing. Thus, invoking (Ust),

$$0 \leq \frac{d}{dh} \mathcal{E}(\varphi + hv) \bigg|_{h=0} \leq -2 \int_a^b u'(t, x)^2 dx \leq 0,$$

so that $u(t) = u(t, a) = u(t, b)$ in (a, b). But $u(t)$ is continuous in x; thus, there exists $\infty \geq l(t) > 0$ such that $S(u(t)) = [0, l(t))$ with $u(t, 0) = tH$ and $u(t, l(t)) = 0$.

We now perform an inner variation in $\mathcal{E}$. Take v be in $\mathcal{C}_0^\infty(0,\infty)$. When $|h|$ is sufficiently small, $x \mapsto \phi_h(x) = x + hv(x)$ is a direct diffeomorphism onto $\mathbb{R}^+$. Moreover, if $\varphi(0) = tH$, the equality also holds for $\varphi_h = \varphi \circ \phi_h^{-1}$, and φ_h converges to φ when h goes to 0. The change of variables $y = \phi_h(x)$ in the energy yields

$$\mathcal{E}(\varphi(t) \circ \phi_h^{-1}) = \int_0^\infty \left(\mu H \frac{u'(t,x)^2}{\phi_h'(x)} + \phi_h'(x)\kappa(2u(t,x)) \right) dx,$$

which in turn leads to

$$\frac{d}{dh}\mathcal{E}(\varphi(t) \circ \phi_h^{-1})\Big|_{h=0} = \int_0^\infty \left(-\mu H u'(t,x)^2 + \kappa(2u(t,x)) \right) v'(x)dx.$$

Thus, invoking (Ust), $\mu H u'(t,x)^2 - \kappa(2u(t,x)) = c$, for some constant c. In particular, $u'(t)$ is continuous on $(0,\infty)$.

Now, take v in $\mathcal{C}_0^\infty(0, l(t))$. Then,

$$\frac{d}{dh}\mathcal{E}(\varphi(t) + hv)\Big|_{h=0} = \int_0^{l(t)} \left(-2\mu H u''(t,x) + 2\kappa'(2u(t,x)) \right) v(x)dx.$$

Thus, invoking (Ust) again, we get that, on $(0, l(t))$, $u'' \geq 0$, that is that u' is monotonically increasing there; since, if $l(t)$ is finite, $u' \equiv 0$ on $(l(t),\infty)$, we conclude that, in any case, u', like u, tends to 0 at infinity, so $c = 0$. From the monotonicity of $u(t)$, we finally get

$$u'(t,x) = -\sqrt{\frac{\kappa(2u(t,x))}{\mu H}}, x > 0, \ u(0) = tH. \qquad (3.15)$$

We then conclude that the solution $\varphi(t) = x + u(t)$ to (Ust) is unique and that it is given by

$$S(\varphi) = [0, l(t)) \quad \text{with} \quad l(t) = \int_0^{tH} \sqrt{\frac{\mu H}{\kappa(2v)}} dv, \qquad (3.16)$$

while

$$\int_{u(t,x)}^{tH} \sqrt{\frac{\mu H}{\kappa(2v)}} dv = x \quad \text{for} \quad x \in S(\varphi). \qquad (3.17)$$

Elementary O.D.E. arguments based on (3.17) would also show that $u(t) \in \mathcal{C}^1([0,\infty); \mathcal{C}^0(0,\infty))$.

The derivation of (3.16), (3.17) only used the monotonicity of κ and its regularity on $(0,\infty)$. In particular, it applies to both the Griffith, and the cohesive setting.

Also note that $l(t)$ and $u(t)$, hence $\varphi(t)$, increase with t, so that irreversibility is automatic, while energy balance is guaranteed by the evoked smoothness of $u(t)$.

This setting offers a striking contrast to that of Subsection 3.1. Here, unilateral stationarity, unilateral local, or unilateral global minimality are indistinguishable, at least for an increasing load.

REMARK 3.8.

a. Of course, a reasonable amount of deception cloaks the analysis. Indeed, we have surreptitiously introduced inner variations in the argument as valid tests for stationarity. This is fine as long as stationarity is understood as including those kind of variations as well. In the presentation of Section 2, stationarity was introduced in the form of a combination of outer and inner variation (see (2.11)). It is in that sense that the re-formulated problem of Proposition 2.1 was equivalent to the original problem (2.1), (2.4) and an investigation of possible additional constraints on that problem resulting from the introduction of inner variations should be undertaken. But "let he who has not sinned throw the first stone".[7]

b. The results are very different when the slab is not homogeneous; a jump in length will occur and it will occur at times which depend on the selected criterion; see (Marigo, 2005).

We end the analysis with a more detailed examination of the explicit form of $u(t)$ for different forms of κ. In Griffith's case,

$$l(t) = tH\sqrt{\frac{\mu H}{k}}, \quad u(t,x) = tH(1 - \frac{x}{l(t)})^+. \tag{3.18}$$

Here, for a given length l of the tear (crack), the total energy at time t is immediately seen to be $\mu H^3 t^2/l + kl$, hence strictly convex in l, so that, according to Proposition 2.4, the smoothness of the evolution $l(t)$ is hardly surprising.

For a Dugdale-type energy, that is a streamlined Barenblatt-type energy of the form $\kappa(\delta) = \min\{\sigma_c\delta; k\}$ first introduced in (Dugdale, 1960), two regions of the jump set $S(\varphi(t))$ should be distinguished: the cohesive zone where $2u(t) < \delta_c := \frac{k}{\sigma_c}$ and the cohesive forces are σ_c; the non-cohesive zone where $2u(t) > \delta_c$ and the cohesive forces vanish. The monotone and continuous character of u imply that both zones are open intervals $(0, \lambda(t))$ and $(\lambda(t), l(t))$, the tip $\lambda(t)$ corresponding to the point where $2u(t) = \delta_c$. For times such that $t \le \delta_c/2H$, there is only a cohesive zone, namely

$$l(t) = \sqrt{\frac{2tH^2\mu}{\sigma_c}}, \quad u(t,x) = tH\left((1 - \frac{x}{l(t)})^+\right)^2.$$

---
[7] John – VIII, 7.

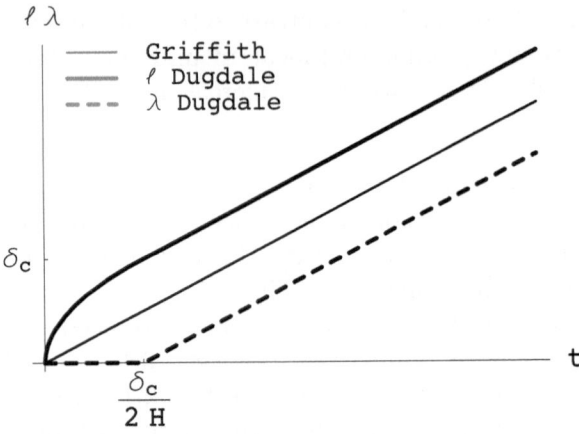

Figure 3.6. A Dugdale type energy – *continuous line*, evolution of the cohesive tip, *dashed line*, evolution of the non-cohesive tip, *thin line*, evolution of the tip of the tearing according to Griffith's model

For $t > \dfrac{\delta_c}{2H}$, both zones coexist, namely,

$$\lambda(t) = \sqrt{\frac{\mu H}{k}}\left(tH - \frac{\delta_c}{2}\right), \quad l(t) = \sqrt{\frac{\mu H}{k}}\left(tH + \frac{\delta_c}{2}\right)$$

and

$$u(t,x) = \begin{cases} tH - \dfrac{x}{\sqrt{\frac{\mu H}{k}}} & \text{in} \quad [0, \lambda(t)] \\[4mm] \dfrac{(l(t) - x)^2}{2\frac{\mu H}{k}\delta_c} & \text{in} \quad [\lambda(t), l(t)]; \end{cases}$$

see Figure 3.6.

Note that, for $t > \delta_c/2H$, the width of the cohesive zone, $l(t) - \lambda(t)$, is independent of t and proportional to δ_c.

REMARK 3.9. If κ is e.g. smooth and $\kappa'(0) = 0$, then the integral $\int_0^\cdot dv/\sqrt{\kappa(2v)}$ diverges. Thus $l(t) \equiv \infty$ for $t > 0$. In other words, initiation is instantaneous and the resulting crack has infinite length! The displacement evolution is still given by (3.17). This remark should be revisited in the light of item e in Remark 4.10 below.

4. Initiation

In this section, we also remain in a 2d setting, but will explicitly mention those results that do not generalize to dimension 3.

Recall the classical example. A semi-infinite 2d homogeneous and isotropic linearly elastic half-plane contains a crack of length l perpendicular to its boundary. In mode I, for a self-equilibrated load of intensity r at ∞, the energy release rate, as computed through Irwin's formula (Irwin, 1958), is proportional to lr^2 and Griffith's criterion consequently requires r to be of the order of $1/\sqrt{l}$ for the crack to move forth. So, as $l \searrow 0$, $r \nearrow \infty$ and no crack will ever appear in the absence of an initial crack.

The mechanics community is of two minds when it comes to crack initiation, or the lack thereof. It claims loud and clear that crack initiation is not within the purview of fracture, because the onset of the cracking process is impurity or imperfection related, yet it relentlessly seeks to predict crack initiation, appealing to extraneous ingredients. Such is one of the motivations of the theory of damage (Lemaître and Chaboche, 1985) which, in essence, substitutes damaged areas for cracks. The thickness of the damaged area – the "process zone" – is controlled by a damage parameter which is in turn assumed to follow an a priori postulated evolution law. A careful tailoring of the damage parameters permits one to control the thickness of the process zone and to collapse it, in the limit, into cracks. The associated phenomenology is however troublesome: What do the damage variables represent and why do they care to follow the postulated evolution laws?

With the refinement of homogenization techniques, modern damage tends to live at a more microscopic scale, the phenomenology being assigned to micro-cracks. The process zone then emerges through averaging and can be tuned in to look like a crack. But then the motion of that "crack" has to be prescribed, and this signals Griffith's return at the macroscopic scale. The reader is invited to reflect upon the validity of introducing damage in any kind of brittle composite such as concrete, where the growth of micro-cracks of the typical inclusion size is controlled by some damage parameter, whereas one should reasonably expect a head-on confrontation with the rather well-defined cracking process. But the result of such a confrontation is predictable because Griffith's classical theory presupposes the presence of the cracks, thus grinding to a halt in such a setting.

And yet how could it be that cracks with length scales of the order of 1/10th of the grain size should obey laws that are unrelated to those that govern cracks with length scales of the order of 10 grain sizes?

Unfettered by ideological bias, we propose to examine the impact of our model upon that issue. Subsection 4.1 investigates the Griffith setting while Subsection 4.2 investigates its cohesive counterpart.

4.1. Initiation – The Griffith case

4.1.1. *Initiation – The Griffith case – Global minimality*

We first address the global minimality setting and recall to that end the weak variational evolution of Subsection 2.5. An important weakness of that evolution is its inability to actually deal with soft devices (at least the kind that were used in the exposition of the formulation). Indeed recalling (2.25) at time 0, it is immediate that, provided that either $f_b(0)$ or $f_s(0)$ are not 0, $\mathcal{E}(0;\cdot)$ has no infimum. For example, if $f_b(0) = 0$ and $f_s(0) \neq 0$ on $\partial_s\Omega$, just take Γ, so as to cut out $\partial_s\Omega$ away from Ω. Sending the excised part to $\pm\infty$ (for a large constant φ on that part) ensures that $-\int_{\partial_s\Omega} f_s(t).\varphi\, ds \searrow -\infty$, while, if $W(0) = 0$, $\int_\Omega W(\nabla\varphi)\, dx = 0$ and $k\mathcal{H}^1(\Gamma)\dot{=}k\mathcal{H}^1(\partial_s\Omega)$.

As illustrated above, this unfortunate byproduct of global minimization cannot be avoided, even if non-interpenetration was accounted for (there is no risk of interpenetration in the example above). A class of body and surface forces for which this does not happen can be evidenced (Dal Maso et al., 2005) in the framework of finite elasticity. But that class does not contain the important case of dead loads (see (Dal Maso et al., 2005), Remark 3.4).

So, in the case of soft devices, the answer to the initiation issue is simple albeit wrong and useless: initiation is immediate as soon as loads are not identically 0 ! The reader should not ridicule the global minimization setting yet. As we shall see later, it yields very reasonable results when hard devices are used, and provides at the least qualitative fits with experimental data in complex settings (see Paragraph 5.1.4).

Consider now the case where $f_b \equiv f_s \equiv 0$. The only load is the boundary deformation $g(t,x))$ which we take to be of the form $tg(x)$. We baptize *proportional loads* these kinds of displacement loadings.

Assume that W is p-homogeneous, $p > 1$. Merely looking at (Ugm), we get that $(\Gamma(t), \varphi(t))$ is a global minimizer for

$$\mathcal{E}(t; \varphi, \Gamma) = \int_\Omega W(\nabla\varphi)\, dx + k\mathcal{H}^1(\Gamma) \qquad (4.1)$$

among all $\overline{\Omega}\backslash\partial_s\Omega \supset \Gamma \supset \Gamma(t)$ and all $\varphi \equiv g(t)$ on $\mathbb{R}^2\backslash\overline{\Omega}$ with $S(\varphi) \subset \Gamma$. So, as long as the body remains purely elastic, $\varphi(t)$ has no jumps outside Γ_0, which we assume closed for simplicity. Then it is the solution of the elastic minimization problem

$$\min_{\varphi}\{\int_{\Omega\backslash\Gamma_0} W(\nabla\varphi)\,dx : \varphi = tg \text{ on } \partial_d\Omega\backslash\Gamma_0\}.$$

Assume that the energy W has the correct functional properties, say smooth, strictly convex and strictly positive for non-zero fields. Note that, although convexity is prohibited by the nonlinear theory of elasticity, it is a natural assumption in the setting of anti-plane shear; it is of course even more acceptable within the linear theory. In any case, the solution $\varphi(t)$ is of the form $t\varphi_g$ with φ_g unique minimizer of (4.1) for $t = 1$. But then, by p-homogeneity,

$$\mathcal{E}(t; \varphi(t), \Gamma_0) = \mathcal{C}t^p + \mathcal{H}^1(\Gamma_0).$$

A competitor to the elastic solution is $\varphi = \begin{cases} 0, & x \in \overline{\Omega} \\ g(t), & \text{otherwise} \end{cases}$, $\Gamma = \Gamma_0 \cup \partial_d\Omega$. For such a test,

$$\mathcal{E}(t; \varphi, \Gamma) = k\mathcal{H}^1(\Gamma_0 \cup \partial_d\Omega).$$

Clearly, if t is large enough, it is energetically favorable to crack (barring exceptional settings where $\mathcal{C} = 0$). We conclude that

PROPOSITION 4.1. *In the global minimality framework, the weak variational evolution for monotonically increasing pure displacement loads will always produce initiation in finite time, provided that the energy is homogeneous.*

We will denote from now on *the initiation time by* t_i, that is the largest time for which $\Gamma(t_i) = \emptyset$. Now that we know that $t_i < \infty$, we would like to understand the circumstances, if any, for which $t_i = 0$. Also, from the global minimality standpoint, the minimum $\Gamma(t)$, $t > t_i$ does not have to satisfy $\mathcal{H}^1(\Gamma(t)) \searrow 0$ as $t \searrow t_i$. We introduce a definition, valid for any $\tau \in [0, T]$.

DEFINITION 4.2. *The crack motion is brutal at time t iff*

$$\lim_{s \searrow t} \mathcal{H}^1(\Gamma(s)) > \mathcal{H}^1(\Gamma(t)).$$

Otherwise the crack motion is progressive at time t.

In other words, the crack motion is brutal at t when the crack experiences a sudden jump in length at that time.

In the case of proportional loads, the singularities of the elastic field lie at the root of the initiation process (see (Francfort and Marigo,

1998), Subsection 4.4). In that analysis it is assumed that W is a quadratic function of the linearized strain, we denote by u_0 the elastic displacement associated with Γ_0 (the maybe empty initial crack) for the load g. We also assume that the field is singular say at only one point x and that, in a neighborhood of that point,

$$u_0(y) = r^\alpha v(\theta) + \hat{u}(y), \ 0 < \alpha < 1,$$

where (r, θ) are polar coordinates with a pole at x. The points x can be thought of as a crack tip (that of Γ_0), or a singular point of the boundary. It is finally assumed that the crack (or add-crack if there is a crack to start with) may only start from x, that the crack (add-crack) $\Gamma(l)$ is a rectifiable curve with a small (but maybe non-zero) length l and that it does not de-bond the domain from $\partial_d\Omega$. We then use an expansion of the bulk energy in terms of the length of a small add-defect; see (Leguillon, 1990). It is given by

$$\mathcal{P}(1, \Gamma_0 \cup \Gamma(l)) = \mathcal{P}(1, \Gamma_0) - \{\mathcal{C}l^{2\alpha} + o(l^{2\alpha})\}, \mathcal{C} > 0, \qquad (4.2)$$

where the potential energy $\mathcal{P}$ is that introduced in (2.21). Let us emphasize that the preceding expansion is formal, so that the argument that we put forth is also formal at this point. A rigorous argument will be outlined a bit later. In view of (4.2), the minimal energy associated with a small (add-)crack of length l, that is

$$\min_{u=tg \text{ on } \partial_d\Omega} \mathcal{E}(t; u, \Gamma_0 \cup \Gamma(l))$$

is

$$t^2\mathcal{P}(1, \Gamma_0) - \mathcal{C}t^2 l^{2\alpha} + k\{l + \mathcal{H}^1(\Gamma_0)\} + t^2 o(l^{2\alpha}), \qquad (4.3)$$

whereas that associated with no (add-)crack is

$$t^2\mathcal{P}(1, \Gamma_0) + k\mathcal{H}^1(\Gamma_0). \qquad (4.4)$$

If $\alpha < \frac{1}{2}$, then for any $t > 0$ a(n) (add-)crack of length less than $\mathcal{C}t^{\frac{2}{1-2\alpha}}$ will carry less energy than no (add)-crack, hence $t_i = 0$ and the crack grows continuously with t, starting with 0 length; this is progressive growth. If $\alpha > \frac{1}{2}$, then denote by $l(t)(\nearrow$ with t) the length of the possible (add-)crack. If $t_i = 0$, then $l(t) \neq 0$ for $t > 0$. But that contradicts the minimality principle if t is very close to 0 because, clearly in such case the expression given by (4.4) is smaller than that given by (4.3). Thus $t_i > 0$. We can apply the same expression (4.3) at time $t = t_i$. Then, if the crack growth is progressive at t_i, $l(t) \searrow 0$ as $t \searrow t_i$. But then $l(t)$, by minimality, must stay 0 in a neighborhood of t_i, a contradiction! This is brutal growth.

Pausing for a moment, we contemplate the implications. The instantaneous creation of a crack of finite length, whether physical or not, is a forbidden feature of the classical theory because it invalidates the very notion of energy release, computed as a derivative. What we witness here is akin to a shock in fluid flow. In truth the necessity of allowing for such events had previously been acknowledged in e.g. (Hashin, 1996), where Rankine-Hugoniot like conditions are suggested, should such a thing happen. In our approach, there is no need to impose additional conditions; they are part of the variational formulation which envisions a much broader collection of test cracks, and, in particular, allows add-cracks with non zero length.

Using similar arguments, we would show that, when $\alpha = \frac{1}{2}$, the crack growth will have a non zero initiation time, and, if there are no singular points, then, either there is no crack growth, or the crack growth will be brutal with a non zero initiation time.

Summing up, we obtain the following

PROPOSITION 4.3. *Assume proportional displacement loads* $g(t) = tg$. *Assume that W is a quadratic function of the linearized strain and that the elastic field is singular say at only one point x, with a singularity in r^α, $0 < \alpha < 1$. Finally assume that the crack (or add-crack) may only start from x, and that the crack (add-crack) $\Gamma(l)$ is a rectifiable curve with a small (but maybe non-zero) length l and that it does not de-bond the domain from $\partial_d\Omega$.*

- *If $\alpha < \frac{1}{2}$ (strong singularity), then $t_i = 0$ and the crack growth is progressive;*

- *If $\alpha > \frac{1}{2}$ (weak singularity), then $t_i > 0$ and the crack growth is brutal;*

- *If $\alpha = \frac{1}{2}$ (critical singularity), then $t_i > 0$;*

- *If there are no singularities, then no crack growth or $t_i > 0$ and the crack growth is brutal.*

The progressive-brutal dichotomy will permeate even the most remote corners of this study. It is one of the cairns that mark the variational approach, in this and other contexts; see e.g. (Francfort and Garroni, 2006).

Proposition 4.3 is encumbered with regularity. The strong or weak variational evolutions do not however presuppose any kind of regularity. This is a delicate analytical issue and the first results in that direction were obtained in (Chambolle et al., 2007). In that 2d study, the cracks are constrained to remain *connected*, a restriction that we have previously mentioned as necessary if the strong formulation is to be retained. Since the arguments in the global minimality setting derive from arguments that only use local minimality, we do not elaborate any further on the analysis at this point, but merely state the obtained results. We will return to this topic in greater details in Paragraph 4.1.2 below.

THEOREM 4.4. *Assume a 2d setting. Assume that*

$$W : \mathbb{R}^2 \mapsto \mathbb{R} \text{ is strictly convex, } C^1$$

$$\gamma |f|^p \leq W(f) \leq \tfrac{1}{\gamma}(|f|^p + 1),\ 1 < p < \infty,\ \text{for some } \gamma > 0,$$

(4.5)

and that the strong variational evolution, with for only load a displacement load $g(t) = tg$, has a solution $\Gamma(t)$ closed, connected, $\varphi(t) \in W^{1,p}(\Omega \backslash \Gamma(t))$.

Call ψ the elastic deformation field associated with g.

– *If, for some $1 < \alpha$,*

$$\sup_{x \in \overline{\Omega}} \left\{ \sup_r \frac{1}{r^\alpha} \int_{B(x;r)} |\nabla \psi|^p\, dx \right\} \leq \mathcal{C},$$

(4.6)

then $t_i > 0$ and the growth is brutal, that is $\mathcal{H}^1(\Gamma(t)) > l^*, t > t_i$, with $l^* > 0$;

– *If $\exists x \in \Omega$ s.t.*

$$\limsup_{r \searrow 0} \frac{1}{r} \int_{B(x;r)} |\nabla \psi|^p\, dx = \infty$$

(4.7)

and on $\overline{\Omega} \backslash \{x\}$, the condition of the first item is satisfied, then $t_i = 0$, the crack starts at $\{x\}$, i.e., $x \in \cap_{t>0}\Gamma(t)$ and

$$\lim_{t \searrow 0} \frac{\mathcal{H}^1(\Gamma(t))}{t} = 0.$$

The same result holds in the setting of linearized elasticity (with $e(u)$ replacing $\nabla \varphi$) for a quadratic energy density.

This theorem is a generalization of Proposition 4.3 to wilder cracks and more general energies. Under suitable regularity assumptions on the load g, existence of a strong variational evolution for connected cracks holds true in 2d. This has been established in (Dal Maso and Toader, 2002) and will be discussed in greater details in Section 5.

We end this paragraph with an analytical example from (Francfort and Marigo, 1998) Section 3.1, which demonstrates brutal initiation (and failure) in the global minimality setting; see Proposition 4.5 below. The reader should note the dependence of the initiation (failure) time t_i upon the length of the strip, an unfortunate byproduct not of the minimization problem, but rather of the presence of a length-scale in Griffith's energy.

Assume that $\Omega = (0, \beta) \times (0, L)$, that the material is linearly isotropic and homogeneous with Young's modulus E, Poisson's ratio ν, and fracture toughness k (see Figure 4.1). Assume also that $\Gamma_0 = \emptyset$, that $u_2 = \sigma_{12} = 0$, $x_2 = 0$ while $u_2 = t$, $\sigma_{12} = 0$, $x_2 = L$, and that $\sigma_{11} = \sigma_{12} = 0$, $x_1 = 0, \beta$.

It is easily seen that the elastic solution $u_e(t), \sigma_e(t)$ is given by

$$
\begin{cases}
u_e(t)(x) = -\left(\nu t \dfrac{x_1}{L} + \mathcal{C}\right) e_1 + t \dfrac{x_2}{L} e_2 \\
\sigma_e(t)(x) = E \dfrac{t}{L} e_2 \otimes e_2.
\end{cases}
$$

The corresponding energy is

$$
\mathcal{E}(t; u_e, \emptyset) = \mathcal{P}(t, \emptyset) = \frac{1}{2} E \frac{t^2}{L} \beta.
$$

Let Γ be an arbitrary crack and denote by $P(\Gamma)$ its projection onto $[0, \beta]$; $P(\Gamma)$ is $\mathcal{H}^1$-measurable because it is compact.

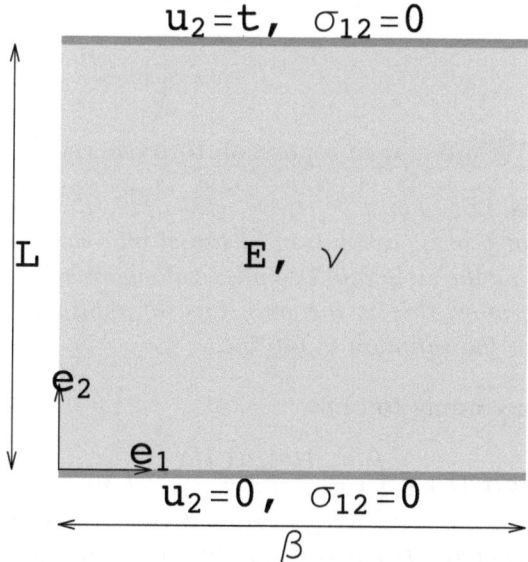

Figure 4.1. Traction of a homogeneous, isotropic cylinder

For an arbitrary crack Γ,

$$\mathcal{P}(t,\Gamma) \geq \left(1 - \frac{\mathcal{H}^1(P(\Gamma))}{\beta}\right)\mathcal{P}(t,\emptyset). \tag{4.8}$$

Indeed, the inequality is obvious if $P(\Gamma) = [0,\beta]$. Otherwise, by quadratic duality,

$$\mathcal{P}(t,\emptyset) \geq \mathcal{P}(t,\Gamma)$$

$$\geq \inf_{v\cdot e_2=0,t \text{ on } x_2=0,L} \frac{1}{2}\int_{\Omega\backslash\Gamma} Ae(v)\cdot e(v)dx$$

$$\geq \inf_{v\cdot e_2=0,t \text{ on } x_2=0,L} \int_{\Omega\backslash\Gamma}\left(\sigma(t)\cdot e(v) - \frac{1}{2}A^{-1}\sigma(t)\cdot\sigma(t)\right)dx, \tag{4.9}$$

where A is the elastic tensor and we choose $\sigma(t) = 0$, if $x \in P(\Gamma) \times (0,L)$ and $\sigma(t) = \sigma_e(t)$ otherwise. Note that the normal vector to the boundary of $P(\Gamma) \times (0,L)$ is e_1, except at $x_2 = 0, L$ where it is e_2; thus it is a statically admissible stress field for the purely elastic problem, as well as for that on $(0,\beta)\backslash P(\Gamma)\times(0,L)$. Then, by elementary application of the divergence theorem,

$$\int_{\Omega\backslash\Gamma}\sigma(t)\cdot e(v)\,dx = \int_{(0,\beta)\backslash P(\Gamma)\times(0,L)}\sigma_e(t)\cdot e(u_e(t))\,dx.$$

Thus,

$$\mathcal{P}(t;\Gamma) \geq \int_{(0,\beta)\backslash P(\Gamma)\times(0,L)}\left(\sigma_e(t)\cdot e(u_e(t)) - \frac{1}{2}A^{-1}\sigma_e(t)\cdot\sigma_e(t)\right)dx$$

$$= \left(1 - \frac{\mathcal{H}^1(P(\Gamma))}{\beta}\right)\mathcal{P}(t,\emptyset)$$

as announced. We are now in a position to prove the following

PROPOSITION 4.5. *For* $t < t_i = \sqrt{2kL/E}$, *the strip* Ω *remains elastic, while for* $t > t_i$, *a solution of the strong variational evolution consists in cutting the strip into two pieces along an arbitrary transverse section. Furthermore this is the only type of solution in the class of cracks for which the infimum is attained.*

Proof. Since, according to (4.8)

$$\mathcal{P}(t;\Gamma) + k\mathcal{H}^1(\Gamma) \geq \left(1 - \frac{\mathcal{H}^1(P(\Gamma))}{\beta}\right)\mathcal{P}(t,\emptyset) + k\mathcal{H}^1(P(\Gamma)),$$

then, provided that $\mathcal{H}^1(P(\Gamma)) \neq 0$ and $\mathcal{P}(t,\emptyset) < k\beta$, the elastic solution is the only global minimizer, which yields the first result in view of the expression for $\mathcal{P}(t, \emptyset)$, except if $\mathcal{H}^1(P(\Gamma)) = 0$, that is except if the crack is parallel to e_2. In that case, (4.8) implies that

$$\mathcal{P}(t; \Gamma) = \mathcal{P}(t, \emptyset),$$

and consequently, $\mathcal{P}(t; \Gamma) + k\mathcal{H}^1(\Gamma) > \mathcal{P}(t, \emptyset)$, unless $\mathcal{H}^1(\Gamma) = 0$.
If $t > t_i$, $\mathcal{P}(t, \emptyset) > k\beta$, so that, according to (4.8),

$$\mathcal{P}(t; \Gamma) + k\mathcal{H}^1(\Gamma) > k\beta,$$

except if $P(\Gamma) = [0, \beta]$, $\mathcal{H}^1(\Gamma) = \beta$, and $\mathcal{P}(t; \Gamma) = 0$. The associated displacement field must then be a rigid body displacement on $\Omega \backslash \Gamma$ which satisfies the boundary displacement conditions at $x_2 = 0, L$, which is impossible unless $\Gamma = [0, \beta] \times \{z\}$, $z \in [0, L]$. $\square$

As mentioned before, initiation and failure coincide in the example above. This will not be the case in most examples.

4.1.2. Initiation – The Griffith case – Local minimality

We now replace the global minimality principle (Ugm) by the local minimality principle (Ulm), and, rather than focus on the strong or weak variational evolutions, merely address initiation in the following sense. Consider a Lipschitz domain Ω and investigate the local minima of

$$\int_{\Omega \backslash \Gamma} W(\nabla\varphi) \, dx + k\mathcal{H}^1(\Gamma), \quad \varphi = g \text{ on } \partial_d\Omega \backslash \Gamma. \tag{4.10}$$

This view of initiation, while apparently completely in agreement with the strong variational evolution, prohibits the pre-existence of a crack because that would contradict the Lipschitz character of the domain. This is because the results exposed below, due to Chambolle, Giacomini and Ponsiglione and found in (Chambolle et al., 2007) stall in the presence of a critical singularity of the elastic field (a $\sqrt{r}$-singularity in the case of quadratic energies).

Those results hinge on the following theorem that we reproduce here without proof, but not without comments, inviting the interested reader to refer to (Chambolle et al., 2007).

THEOREM 4.6. *Assume a 2d setting. If W satisfies (4.5), and ψ, the elastic solution, satisfies (4.6) (the singularities are uniformly weaker than the critical singularity), then, $\exists l^* > 0$ s.t., for all connected Γ, closed in $\overline{\Omega}$, with $\mathcal{H}^1(\Gamma) < l^*$, and all $\varphi \in W^{1,p}(\Omega \backslash \Gamma)$ with $\varphi = g$ on $\partial_d\Omega \backslash \Gamma$,*

$$\int_\Omega W(\nabla\psi)dx < \int_{\Omega(\backslash\Gamma)} W(\nabla\varphi) \, dx + k\mathcal{H}^1(\Gamma).$$

Note that it matters not whether the integral on the right is taken over Ω or $\Omega \backslash \Gamma$.

Because of the growth assumption contained in (4.5), (4.6) may be viewed as stating that (uniformly in x) the bulk energy on a small disk is energetically more favorable than the surface energy associated with a crack along the diameter of that disk. Then, the conclusion is that the elastic response to the load g is energetically better than that associated with a connected crack, if that length is less than l^*. In the case of a quadratic energy and in the terminology used in Proposition 4.3, this is saying that, in the case of uniform weak singularities, crack initiation can only be brutal.

As an immediate corollary of the theorem above, we obtain the following local minimality result.

COROLLARY 4.7. *In the setting of Theorem 4.6, the elastic solution ψ is a local minimum for (4.10) for the L^1–distance.*

Proof. Take Γ_n, $\varphi_n \in W^{1,p}(\Omega \backslash \Gamma_n)$ with $\|\varphi_n - \psi\|_{L^1(\Omega)} \searrow 0$ and assume that

$$\int_\Omega W(\nabla \psi) dx > \int_\Omega W(\nabla \varphi_n) \, dx + k \mathcal{H}^1(\Gamma_n).$$

By a lower semi-continuity result due to Ambrosio (see (Ambrosio, 1994), or Theorem D in the Appendix),

$$\int_\Omega W(\nabla \psi) dx \leq \liminf_n \int_\Omega W(\nabla \varphi_n) \, dx, \qquad (4.11)$$

which is impossible unless $\liminf_n \mathcal{H}^1(\Gamma_n) = 0$. But, then, for a subsequence $\{k(n)\}$ of $\{n\}$,

$$\mathcal{H}^1(\Gamma_{k(n)}) \leq l^*,$$

in which case, according to Theorem 4.6,

$$\int_\Omega W(\nabla \psi) dx < \int_\Omega W(\nabla \varphi_{k(n)}) \, dx + k \mathcal{H}^1(\Gamma_{k(n)}),$$

in which case, according to Theorem 4.6,

in contradiction with the starting assumption.   □

The strong L^1-topology is least intrusive in terms of locality, in the sense that closeness in that topology will be implied by closeness in any reasonable topology. The corollary thus states that – modulo the connectedness restriction – the elastic solution is always a local minimizer whenever the associated field exhibits at most weak singularities. In this respect, local minimality is closer to the classical theory than global minimality because, as partially discussed at the onset of the section, the classical theory cannot initiate a crack without a pre-crack (a critical singularity), or a strong singularity.

REMARK 4.8. Here, the energy release rate associated with the elastic solution is a meaningful notion. Indeed, take any connected crack $\Gamma(l)$ of length l; then, for any fracture toughness k', and provided that l is small enough,

$$\int_{\Omega} W(\nabla\psi)dx < \int_{\Omega(\backslash\Gamma(l))} W(\nabla\varphi(l)) \, dx + k'l,$$

where $\varphi(l)$ is the elastic solution on $\Omega\backslash\Gamma(l)$. Hence

$$\limsup_{l\searrow 0} \left\{ \int_{\Omega} W(\nabla\psi)dx - \int_{\Omega(\backslash\Gamma(l))} W(\nabla\varphi(l)) \, dx \right\} \leq 0.$$

But the quantity above is always non-negative and we conclude that the limsup is a limit and that limit is 0. In other words, the energy release rate in such a setting is 0, which comforts the intuition provided by the classical theory.

The case of a strong singularity, *i.e.*, of points x s.t.

$$\limsup_{r\searrow 0} \frac{1}{r} \int_{B(x,r)} |\nabla\psi|^p \, dx = \infty,$$

is easier to handle, because, clearly, in such a case, it is energetically more advantageous to replace ψ in a small ball around x by 0.

All results quoted in this paragraph extend to the vectorial case (plane hyperelasticity) and to the setting of linearized elasticity (Chambolle et al., 2007).

The classically trained mechanician will sigh in relief: no initiation with local minimality; the elastic solution remains meta-stable as it should. "And then this 'should' is like a spendthrift sigh, That hurts by easing"[8]. Indeed, the lack of uniqueness may still produce crack initiation as the load increases. In other words, the only conclusion to draw from the preceding analysis is that a departure from the elastic solution will necessitate the nucleation of a crack of non-zero length. But the energetic barrier might be arbitrarily small, much smaller than that for which such a nucleation may be rejected as unphysical.

As will be seen in the next subsection, the introduction of a cohesive energy leads to a very different panorama.

---

[8] Shakespeare – Hamlet – IV, 7

4.2. INITIATION – THE COHESIVE CASE

Griffith's indictment usually mentions unbounded stresses as a prime culprit. Indeed, stress singularity is a by-product of the absence of cohesiveness, at least in a linear setting, because the elastic solution on the uncracked part of the domain must then blow up near the crack tip. Barenblatt then suggests local cohesiveness near the crack tip as a correcting term preventing stress singularities. But he falls short of addressing the impact of cohesiveness on initiation.

Del Piero (Del Piero, 1997) is, to our knowledge, first in his attempt to include cohesiveness in a variational approach of crack initiation. His approach, which is energy based and one-dimensional merges with ours in that setting. In this respect, the analysis presented in Paragraph 3.1.4 is essentially his. We acknowledge it now so as to emphasize his contributions to cohesive initiation.

Initiation in the cohesive setting is more easily grasped as a local minimality issue, because, as will be seen below, global minimality in that setting entails relaxation, whereas local minimality may not. Actually, reneging on earlier commitments, we do not even resort to local minimality. Our argument is based solely on the use of unilateral stationarity, hence, according to the arguments put forth in Section 2, completely in agreement with Griffith's view of fracture, albeit for a Barenblatt type energy. The treatment of irreversibility in Section 5 will mirror this and also use unilateral stationarity.

Consequently, we first address the stationarity issue. In a first paragraph the 1d problem is thoroughly dissected. Higher dimensional settings are discussed in the following paragraph. Finally, global minimality and its link to relaxation will be the topic of the last paragraph.

4.2.1. *Initiation – The cohesive 1d case – Stationarity*

Consider a homogeneous bar of length L, clamped at $x = 0$, subject to a load f_b along its length and to a force f_s at $x = L$. The deformation map is $\varphi(x)$, with possible jumps $S(\varphi) \in [0, L]$. As already mentioned, non-interpenetration is much easier to handle when dealing with cohesive models, so that, we impose non-negative jumps for φ. The analysis follows closely that developed in Paragraph 3.1.3. The ambient space for probing initiation is (roughly)

$$\mathcal{S} := \{\varphi \in SBV(\mathbb{R}) : S(\varphi) \subset [0, L), \varphi \equiv 0 \text{ on } (-\infty, 0); [\varphi] \geq 0 \text{ on } S(\varphi)\}.$$

The work of the external loads is given by

$$\mathcal{F}(\varphi) := \int_0^L f_b(x)\varphi(x) \, dx + f_s\varphi(L).$$

We assume a strictly convex energy $W(F)$. The "surface" energy density $\kappa(\delta)$ is defined for $\delta \geq 0$, with $\kappa(0) = 0$. It is $\mathcal{C}^1$ and σ_c, its right-derivative at 0, is the maximal value of the derivative κ'. The total energy of the bar is

$$\mathcal{E}(\varphi) = \int_0^L W(\varphi') \, dx + \sum_{S(\varphi)} \kappa([\varphi]) - \mathcal{F}(\varphi).$$

Denote by σ_e the stress field associated with its elastic response φ_e, that is that with no jumps. Then,

$$\varphi_e(0) = 0, \quad \varphi_e' = W'^{-1}(\sigma_e), \quad \sigma_e' + f_b = 0 \text{ in } (0, L), \quad \sigma_e(L) = f_s.$$

Note that φ_e and σ_e are uniquely determined.

Assume that φ_e satisfies (Ust); then,

$$\frac{d}{d\varepsilon}\mathcal{E}(\varphi_e + \varepsilon\zeta)\Big|_{\varepsilon=0} \geq 0, \quad \forall \zeta \text{ such that } \begin{cases} \zeta(0-) = 0 \\ [\zeta] > 0 \text{ on } S(\zeta). \end{cases} \tag{4.12}$$

But

$$\begin{aligned} \frac{d}{d\varepsilon}\mathcal{E}(\varphi_e + \varepsilon\zeta)\Big|_{\varepsilon=0} &= \int_{(0,L)} \sigma_e \zeta' \, dx + \sum_{S(\zeta)} \sigma_c[\zeta] - \mathcal{F}(\zeta) \\ &= \sum_{x \in S(\zeta)} (\sigma_c - \sigma_e)[\zeta(x)]. \end{aligned}$$

Thus, if (4.12) is satisfied, then

$$\sup_{x \in [0,L]} \sigma_e(x) \leq \sigma_c.$$

The elastic stress must be everywhere smaller than the critical stress σ_c.

The elastic solution may not be the only unilateral stationarity point. Thus, assume that φ satisfies unilateral stationarity; here unilateral sationarity means

$$\frac{d}{d\varepsilon}\mathcal{E}(\varphi + \varepsilon\zeta)\Big|_{\varepsilon=0} \geq 0, \tag{4.13}$$

for all ζ's such that $\zeta(0-) = 0$ and $[\zeta] > 0$ on $S(\zeta)\backslash S(\varphi)$. Note that $[\zeta]$ may be arbitrary on $S(\varphi)$ since $[\varphi + \varepsilon\zeta] \geq 0$ for ε small enough. Then, upon setting $\sigma = W'(\varphi')$, we get, for any ζ with $[\zeta] > 0$ on $S(\zeta)\backslash S(\varphi)$,

$$0 \leq \int_{(0,L)} \sigma \zeta' \, dx - \mathcal{F}(\zeta) + \sum_{S(\zeta)\backslash S(\varphi)} \sigma_c[\zeta] + \sum_{S(\varphi)} \kappa'([\varphi])[\zeta].$$

Smooth test functions yield

$$\begin{cases} \sigma' + f_b = 0 \text{ in } (0, L) \\ \sigma(L) = f_s, \end{cases}$$

so that σ is continuous on $[0, L]$, and non-smooth test functions yield in turn

$$0 \le \sum_{S(\zeta)\backslash S(\varphi)} (\sigma_c - \sigma)[\zeta] + \sum_{S(\varphi)} (\kappa'([\varphi]) - \sigma)[\zeta].$$

Thus, for all $x \in [0, L]$,

$$\sigma(x) \le \sigma_c \text{ in } [0, L]\backslash S(\varphi), \quad \sigma(x) = \kappa'([\varphi](x)) \text{ on } S(\varphi). \qquad (4.14)$$

Since σ is continuous and $S(\varphi)$ at most countable, the first condition in (4.14) forces

$$\sigma(x) \le \sigma_c, \forall x \in [0, L],$$

whereas the second condition links the (normal) stress at a discontinuity point to the cohesive force – the derivative of the surface energy – at that point.

We have established the following

PROPOSITION 4.9 (1d - cohesive initiation). *In the cohesive 1d context, whether starting from the elastic solution or from an already discontinuous solution, initiation – that is the non-stationarity of the solution – will occur if the stress field at any point becomes greater than the critical stress, defined as the slope at 0 of the cohesive surface energy.*

This calls for several remarks.

REMARK 4.10.

a. The initiation condition in the above proposition is obtained without invoking any kind of minimality criterion; it is thus directly in the spirit of Griffith's formulation;

b. The finiteness of the right derivative of the surface energy density is the essential feature that allows to adjudicate initiation in terms of a critical stress criterion. In the setting of a Griffith's type surface energy, that derivative is infinite ($\sigma_c = +\infty$ if you will) and the criterion is moot;

c. The other features of the surface energy are irrelevant to the issue of initiation;

d. There is no lower bound on σ, or, in other words, compressive stresses can be as large as they wish, as expected from the condition of non-interpenetration;

e. The slope of the surface energy at 0 *cannot be* 0. Indeed, it would then be impossible to find a solution to unilateral stationarity (4.12), (4.13) with a countable number of discontinuity points, lest the critical stress criterion be violated away from those points.

This apparently innocuous observation delivers, in our opinion, a devastating blow to a whole slew of models – especially popular among numerical mechanicians – that propose to tackle fracture through the introduction of surface energies with 0-slopes; see (Chaboche et al., 2001), (Tvergaard, 1990). In view of the above, such models are doomed.

f. Haziness seems to surround the true nature of initiation, that is when $\sigma(x) > \sigma_c$ for some point $x \in [0, L]$. Clearly, in view of (4.14), stationarity will not be met anymore. We conjecture that, at such a time, dynamical effects will upstage our usual variational partners but, in truth, this amounts to little more than hearsay at this point.

The 1d result may be generalized to a multi-dimensional setting. This is the object of the following paragraph.

4.2.2. *Initiation – The cohesive 3d case – Stationarity*

The setting and notation are those of Section 2, adapted to 3d. We further assume, for simplicity, homogeneity and isotropy of the material properties and denote by φ_e the – or at least an – elastic response and by σ_e the associated stress field. Thus,

$$\sigma_e = \frac{\partial W}{\partial F}(\nabla \varphi_e), \quad \text{div } \sigma_e + f_b = 0 \text{ in } \Omega, \quad \sigma_e n = f_s \text{ on } \partial_s \Omega. \quad (4.15)$$

We also impose the following *a priori* regularity on σ_e:

$$\sigma_e \text{ is the restriction to } \Omega \text{ of an element of } \mathcal{C}^0(\mathbb{R}^3; \mathbb{R}^3). \quad (4.16)$$

The surface energy density is a function Φ of both the jump ψ and the orientation ν at each point of the discontinuity set. Isotropy requires that

$$\Phi(Q\nu, Q\psi) = \Phi(\nu, \psi), \quad \forall Q \in SO^3, \forall \nu \in \mathcal{S}^2, \forall \psi \in \mathbb{R}^3.$$

Hence Φ is a function of the invariants of the 2×3-matrix (ν, ψ) (Ciarlet, 1986). But ν is a unit vector so that the only invariants are $\psi \cdot \nu$ and $|\psi|$, or equivalently, $\psi \cdot \nu$ and $|\psi - \psi \cdot \nu \, \nu|$. Further, non-interpenetration demands that $\psi \cdot \nu \geq 0$. Define κ on $[0, \infty)^2$ such that

$$\Phi(\nu, \psi) = \kappa(\psi \cdot \nu, |\psi - \psi \cdot \nu \, \nu|), \quad \forall \nu \in \mathcal{S}^2, \forall \psi \in \mathbb{R}^3 : \psi \cdot \nu \geq 0.$$

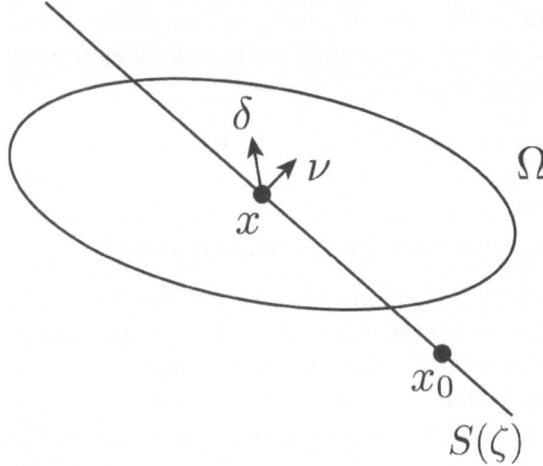

Figure 4.2. Test jump set

We further impose that $\kappa(0,0) = 0$, $\kappa \geq 0$, that κ be continuous, and also that κ be directionally differentiable at 0, that is that

(DirD) There exists a (positively 1-homogeneous) function κ_0 such that $0 < \kappa_0(\alpha, \beta) = \lim_{h \to 0+} 1/h \ \kappa(h\alpha, h\beta)$. In particular, σ_c and τ_c respectively denote $\kappa_0(1, 0) > 0$ and $\kappa_0(0, 1) > 0$.

When κ is differentiable at $(0,0)$, then κ_0 is linear and $\kappa_0(\alpha, \beta) = \sigma_c \alpha + \tau_c \beta$.

If the field φ_e is a unilateral stationary point, then, for any $\overline{\zeta} \geq 0 \in \mathcal{C}_0^\infty(\Omega)$ and for any $\nu, \delta \in \mathcal{S}^2$ with $\delta \cdot \nu \geq 0$, take $\zeta = \overline{\zeta} \chi_{\{x:(x-x_0).\nu \geq 0\}} \delta$ (so that $[\zeta] \cdot \nu \geq 0$ on $S(\zeta) \subset \{x : \nu.(x - x_0) = 0\}$; see Figure 4.2).

We obtain

$$\frac{d}{d\varepsilon} \mathcal{E}(\varphi_e + \varepsilon\zeta)\Big|_{\varepsilon=0+} = \int_\Omega \sigma_e \cdot \nabla\zeta \ dx - \mathcal{F}(\zeta)$$
$$+ \int_{S(\zeta)\backslash\partial_s\Omega} \kappa_0\Big([\zeta]\cdot\nu, |[\zeta] - [\zeta]\cdot\nu \ \nu|\Big) \ d\mathcal{H}^2,$$

which in view of (4.16), (4.15), yields

$$0 \leq \int_{S(\zeta)} \Big(\kappa_0\big([\zeta]\cdot\nu, |[\zeta] - [\zeta]\cdot\nu \ \nu|\big) - \sigma_e\nu\cdot[\zeta]\Big) \ d\mathcal{H}^2.$$

Because $\overline{\zeta}$ is arbitrary and κ_0 1-homogeneous, this yields, $\mathcal{H}^2$-a.e. on $\{x \in \Omega : (x - x_0).\nu = 0\}$,

$$\kappa_0(\delta\cdot\nu, |\delta - \delta\cdot\nu \ \nu|) \geq \sigma_e(x)\nu\cdot\delta.$$

Varying x_0 and recalling assumption (4.16), we conclude that, at least when the elastic solution is unique,

PROPOSITION 4.11 (3d-cohesive initiation). *In the cohesive 3d context and under assumption (4.16), initiation starting from the elastic solution will occur if the stress field at any point is such that there exists $\nu \in S^2, \psi \in \mathbb{R}^3$ – with $\psi \cdot \nu \geq 0$ – satisfying*

$$\sigma_e \nu \cdot \psi > \kappa_0(\psi \cdot \nu, |\psi - \psi \cdot \nu \, \nu|).$$

We call this condition the yield stress condition.

As in the 1d case, the yield stress condition is not specific to the elastic state, but rather it appears as a yield stress condition for any state. The interested reader is invited to consult (Charlotte et al., 2006) for a proof.

Such a simple trove hides many riches, which we unwrap in the following two remarks.

REMARK 4.12. Whenever the surface energy density κ is differentiable at the origin, the yield stress criterion becomes

$$\sigma\nu \cdot \psi > \sigma_c \psi \cdot \nu + \tau_c |\psi - \psi \cdot \nu \, \nu|,$$

for some pair (ν, ψ) of unit vectors with $\psi \cdot \nu \geq 0$. Let τ be a unit vector orthogonal to ν. Decomposing ψ into its normal and tangential part:

$$\psi = \cos \theta \, \nu + \sin \theta \, \tau, \quad -\pi/2 \leq \theta \leq \pi/2,$$

we then get, for some $(\nu, \tau) \in S^2 \times S^2$ with $\tau \cdot \nu = 0$,

$$(\sigma\nu \cdot \nu - \sigma_c) \cos \theta + \sigma\nu \cdot \tau \sin \theta - \tau_c |\sin \theta| > 0,$$

or, equivalently,

$$\max_{\nu \in S^2} \sigma\nu \cdot \nu > \sigma_c, \quad \text{or} \quad \max_{(\nu,\tau) \in S^2 \times S^2 \, : \, \nu \cdot \tau = 0} \sigma\nu \cdot \tau > \tau_c. \tag{4.17}$$

This means that the stress field σ violates the *criterion of maximal traction* or *maximal shear*. Whenever σ is symmetric (in linearized elasticity for example) these criteria can be written solely in terms of the eigenvalues $(\sigma_1, \sigma_2, \sigma_3)$ of the stress tensor. They read as

$$\max_i \sigma_i > \sigma_c, \quad \text{or} \quad \max_{i,j}(\sigma_i - \sigma_j) > 2\tau_c.$$

REMARK 4.13. The initiation criteria of maximal traction or maximal shear – see Remark 4.12 – assume isotropy and differentiability of the surface energy density. In this long remark, we relax the differentiability condition and assume an isotropic directionally differentiable surface energy density. We decompose the stress vector $\sigma\nu$ into its normal and tangential parts

$$\sigma\nu = \Sigma\nu + T\tau, \quad \tau \in \mathcal{S}^2 \text{ with } \tau \cdot \nu = 0.$$

Then, the yield stress criterion is satisfied if and only if the stress vector (Σ, T) lies outside the following convex set of the Mohr diagram, that is

$$\Sigma > \sigma_c \text{ or } |T| > \kappa_\star(\Sigma) \tag{4.18}$$

with

$$\kappa_\star(\Sigma) = \inf_{\lambda \geq 0} \{\kappa_0(\lambda, 1) - \lambda\Sigma\}.$$

The function $\kappa_\star$ – which gives rise to the so-called *intrinsic curve* $|T| = \kappa_\star(\Sigma)$ – is such that

a. The function $\kappa_\star$ is defined for $\Sigma \in (-\infty, \sigma_c)$, concave, continuous, decreasing and $\lim_{\Sigma \to -\infty} \kappa_\star(\Sigma) = \tau_c = \kappa_0(0, 1)$. $\kappa_\star$ is non negative for $\Sigma \in (-\infty, \sigma_c^\star]$ with $\sigma_c^\star \leq \sigma_c = \kappa_0(1, 0)$; consequently,

b. The domain of the admissible (Σ, T) delimited by the intrinsic curve is convex, symmetric with respect to the axis $T = 0$, unbounded in the direction of negative normal stress and bounded by $\sigma_c^\star$ in the direction of positive normal stress, see Figure 4.3.

To see this, we decompose $\psi \in \mathcal{S}^2$ with $\psi \cdot \nu \geq 0$ into its normal and tangential parts

$$\psi = \cos\theta\nu + \sin\theta\tau', \quad \tau' \in \mathcal{S}^2 : \tau' \cdot \nu = 0, \quad \theta \in [0, \pi/2].$$

Then, the yield stress criterion reads as: $\exists\theta \in [0, \pi/2]$, and $\exists\tau' \in \mathcal{S}^2$ with $\tau' \cdot \nu = 0$ such that

$$T\tau \cdot \tau' \sin\theta > \kappa_0(\cos\theta, \sin\theta) - \Sigma\cos\theta.$$

This will happen provided that, for some $\theta \in [0, \pi/2]$,

$$|T|\sin\theta > \kappa_0(\cos\theta, \sin\theta) - \Sigma\cos\theta.$$

If $\theta = 0$, this gives $\Sigma > \kappa_0(1, 0) = \sigma_c$. If $\theta \neq 0$, then $|T| > \kappa_0(\lambda, 1) - \lambda\Sigma$ for some $\lambda \geq 0$. Condition (4.18) follows.

We now investigate the properties of $\kappa_\star$. Define

$$\bar{\kappa}_0(\lambda) = \begin{cases} \kappa_0(\lambda, 1) & \text{if } \lambda \geq 0 \\ +\infty & \text{otherwise.} \end{cases}$$

Then $\kappa_\star = -\bar{\kappa}_0^\star$, the Legendre transform of $\bar{\kappa}_0$, which proves that $\kappa_\star$ is concave and continuous. If $\Sigma > \sigma_c$, take $\lambda_n = n$; then $\kappa_\star(\Sigma) \leq n(\kappa_0(1, 1/n) - \Sigma)$. Since $\lim_{n\to\infty} \kappa_0(1, 1/n) = \sigma_c$, $\kappa_\star(\Sigma) = -\infty$. If $\Sigma < \sigma_c$, then $\lambda \mapsto \kappa_0(\lambda, 1) - \lambda\Sigma$ is continuous and tends to $+\infty$ as λ tends to $+\infty$. Thus the infimum is reached (and finite).

To prove that $\kappa_\star$ is decreasing, consider $\Sigma_1 < \Sigma_2 < \sigma_c$ and let λ_1 and λ_2 be points where the infimum is reached. Then

$$\kappa_\star(\Sigma_1) = \kappa_0(\lambda_1, 1) - \lambda_1 \Sigma_1,$$

while

$$\kappa_\star(\Sigma_2) = \kappa_0(\lambda_2, 1) - \lambda_2 \Sigma_2 \leq \kappa_0(\lambda_1, 1) - \lambda_1 \Sigma_2,$$

hence $\kappa_\star(\Sigma_1) - \kappa_\star(\Sigma_2) \geq (\Sigma_2 - \Sigma_1)\lambda_1 \geq 0$.

To prove that $\lim_{\Sigma \to -\infty} \kappa_\star(\Sigma) = \tau_c$, first note that $\kappa_\star(\Sigma) \leq \kappa_0(0, 1) - 0 \cdot \Sigma = \tau_c$ for all Σ. Then take $\Sigma_n = -n$ and let λ_n be the associated sequence of minimizers. In turn, $\kappa_\star(-n) = \kappa_0(\lambda_n, 1) + n\lambda_n \leq \tau_c$, and, since κ_0 is positive, $\lim_{n\to\infty} \lambda_n = 0$. Consequently, since $\tau_c \geq \kappa_\star(-n) \geq \kappa_0(\lambda_n, 1)$, we get $\lim_{n\to\infty} \kappa_\star(-n) = \tau_c$.

The function $\kappa_\star$ may be negative at $\Sigma = \sigma_c$; see the example below. If $\kappa_\star(\sigma_c) \geq 0$, then the domain of admissible stress vectors is $\{(\Sigma, T) : -\infty < \Sigma \leq \sigma_c, |T| \leq \kappa_\star(\Sigma)\}$. Otherwise, by continuity and monotonicity, there exists $\sigma_c^\star$ such that $\kappa_\star(\sigma_c^\star) = 0$ and $\kappa_\star(\Sigma) < 0, \forall \Sigma > \sigma_c^\star$. In that case, the domain of admissible stress vectors is $\{(\Sigma, T) : -\infty < \Sigma \leq \sigma_c^\star, |T| \leq \kappa_\star(\Sigma)\}$.

In the case where κ_0 is linear, we obtain

$$\kappa_\star(\Sigma) = \begin{cases} \tau_c & \text{if } \Sigma \leq \sigma_c \\ -\infty & \text{otherwise} \end{cases}$$

and recover the criteria (4.17) of maximal traction and maximal shear.

All of the above holds true for a fixed vector ν. We must now vary ν along the unit sphere $\mathcal{S}^2$. Assume once again symmetry of the stress field σ, and denote by $\sigma_1 \leq \sigma_2 \leq \sigma_3$ the ordered eigenvalues of σ. The point (Σ, T) spans the domain bounded by the three Mohr circles. So, σ will satisfy the yield stress criterion if and only if either $\sigma_3 > \sigma_c$, or the greatest Mohr circle reaches outside the convex hull of all greatest Mohr circles lying inside the domain bounded by the intrinsic curve, that is

$$\frac{\sigma_3 - \sigma_1}{2} > \bar{\kappa}_\star \left(\frac{\sigma_1 + \sigma_3}{2}\right),$$

where $\bar{\kappa}_\star(s) := \inf_{0 \leq \theta \leq \pi/2} \{\kappa_0(\cos\theta, \sin\theta) - s\cos\theta\}$.

As in 1d, the asymmetric behavior between traction and compression is a byproduct of the non-interpenetration condition. The convexity of

66   B. Bourdin, G. Francfort and J.-J. Marigo

the domain of admissible stress tensors is a direct consequence of the stationarity condition. That it should be obtained from an intrinsic curve in the Mohr diagram – and consequently that it does not depend upon the intermediary stress eigenvalue σ_2 – is a consequence of both stationarity and isotropy. As such, this may be no longer apply when anisotropy is considered.

We end the reader's initiation with three examples that illustrate Remark 4.13. Here, κ is taken to be

$$\kappa(\alpha, \beta) = k \left(1 - \exp\left(-\frac{\kappa_0(\alpha, \beta)}{k} \right) \right),$$

with κ_0 positive, continuous and one-homogeneous. Then κ is not Fréchet differentiable at $(0,0)$, but its directional derivative is just κ_0.

If κ_0 is convex, say for example

$$\kappa_0(\alpha, \beta) = 2\sqrt{\sigma_c^2 \alpha^2 + \tau_c^2 \beta^2} - \sigma_c \alpha - \tau_c \beta,$$

then, a straightforward computation leads to

$$\kappa_\star(\Sigma) = \begin{cases} \tau_c & \text{if } \Sigma \leq -\sigma_c \\ \tau_c \left(\sqrt{4 - \left(1 + \dfrac{\Sigma}{\sigma_c}\right)^2} - 1 \right) & \text{if } -\sigma_c \leq \Sigma \leq \sigma_c. \end{cases}$$

Thus, $\kappa_\star$ is positive if and only if $\Sigma \leq (\sqrt{3} - 1)\sigma_c \equiv \sigma_c^\star$ and $\sigma_c^\star$ is the maximal traction that the material can sustain. The intrinsic curve is represented on Figure 4.3.

If κ_0 is concave, then the minimum of $\kappa_0(\lambda, 1) - \lambda\Sigma$ is reached at $\lambda = 0$ (since, as previously observed, that infimum is reached at a finite

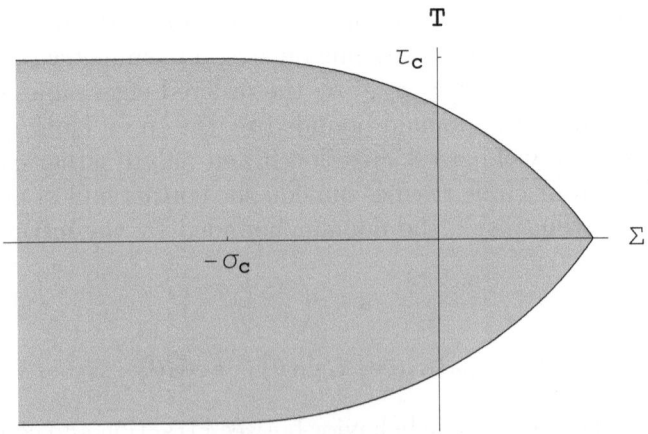

Figure 4.3. The set of the admissible stress vectors in the Mohr diagram

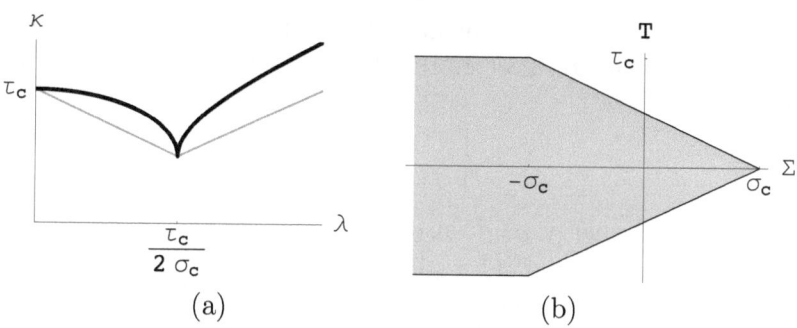

Figure 4.4. (a) the graphs of the function $\lambda \mapsto \kappa_0(\lambda, 1)$ and of its convex envelop $\lambda \mapsto \kappa_0^{\star\star}(\lambda, 1)$. (b) the corresponding intrinsic curve

λ) and we recover the maximal traction and the maximal shear criteria, as in the case of a linear κ_0.

If finally κ_0 is neither concave nor convex, say for example

$$\kappa_0(\alpha, \beta) = \tau_c \frac{\beta}{2} + \sqrt{\left| \sigma_c^2 \alpha^2 - \tau_c^2 \frac{\beta^2}{4} \right|},$$

then $\kappa_0(\lambda, 1)$ is neither convex nor concave, and its convex envelop $\lambda \mapsto \kappa_0^{\star\star}(\lambda, 1)$ is made of two line segments, see Figure 4.4. Then the minimization of $\lambda \mapsto \kappa_0(\lambda, 1) - \lambda\Sigma$ on $[0, \infty)$ is equivalent to the minimization of its convex envelope $\lambda \mapsto \kappa_0^{\star\star}(\lambda, 1) - \lambda\Sigma$, see (Dacorogna, 1989). Then

$$\kappa_\star(\Sigma) = \begin{cases} \tau_c & \text{if } \Sigma \leq -\sigma_c \\ \dfrac{\tau_c}{2}\left(1 - \dfrac{\Sigma}{\sigma_c}\right) & \text{if } |\Sigma| \leq \sigma_c, \end{cases}$$

$\sigma_c^\star = \sigma_c$ and the domain bounded by the intrinsic curve is represented on Figure 4.4. When considering the envelop of the greatest Mohr circle that lies inside that domain, the corners $(-\sigma_c, \pm\tau_c)$ disappear and finally the domain of the admissible stress tensors is given by

$$\bar{\kappa}_\star(\Sigma) = \begin{cases} \tau_c & \text{if } \Sigma \leq \sigma_c - \sqrt{4\sigma_c^2 + \tau_c^2} \\ \tau_c \dfrac{\sigma_c - \Sigma}{\sqrt{4\sigma_c^2 + \tau_c^2}} & \text{if } \sigma_c - \sqrt{4\sigma_c^2 + \tau_c^2} \leq \Sigma \leq \sigma_c. \end{cases}$$

4.2.3. *Initiation – The cohesive case – Global minimality*

We now address the global minimality setting. It was intimated in Subsection 2.6 that the weak variational evolution was ill-posed even at the initial time, and that the minimization problem at $t = 0$ had to be relaxed. This is easily understood through a simple energetic comparison. Assume, in e.g. 1d, that, on $(0, L)$, the gradient of the field φ is

$$\frac{d\varphi}{dx} = \begin{cases} f, \ 0 < x \leq a < L \\ f + g, \ a < x < L. \end{cases}$$

Then the total energy paid on $(0, L)$ is the bulk energy paid on that interval, i.e., $W(f)a + W(f + g)(L - a)$. Between a and L, $\varphi - fx$ has increased by $g(L - a)$. Thus take instead n small jumps of amplitude $g(L - a)/n$ for $x \in (a, L)$. The associated energy is then $n\kappa(g(L - a)/n)$; as n becomes large the latter goes to $\sigma_c g(L - a)$ (recall that $\kappa'(0) := \sigma_c$). Consequently, the total energy paid is then $W(f)L + \sigma_c g(L - a)$.

Comparing both energies reduces to an investigation of the sign of

$$(W(f + g) - W(f)) - \sigma_c g.$$

Letting g go to 0, we get that the first energy is smaller than the second one iff

$$W'(f) \leq \kappa'(0) = \sigma_c.$$

So, for gradients larger than $(W')^{-1}(\sigma_c)$, it is energetically more favorable to use jumps. The relaxation will thus truncate the bulk energy at that level and replace it with an linearly growing energy (recall that we are assuming throughout that W has p–growth with $p > 1$) as announced in Subsection 2.6.

Now, the a priori bounds on the minimizing sequences for

$$\int_0^L W(\frac{d\varphi}{dx}) \, dx + \sum_{x \in S(\varphi)} \kappa([\varphi(x)]),$$

do not permit application of Ambrosio's compactness theorem (2.24), mainly because

$$\lim_{t \searrow 0} \frac{\kappa(t)}{t} = \sigma_c \neq \infty,$$

and the limit field might thus live, not in $SBV(0, L)$, but only in $BV(0, L)$. Indeed, any Cantor function can be approximated by e.g. SBV-functions with very small jumps only; see (Ambrosio et al., 2000), Section 3.2. This explains the appearance of the term $\sigma_c C(\varphi)$ in the relaxed functional (2.30). In other words, the original energy can promote through micro-cracking some kind of energetically charged "diffuse crack" with zero bulk and surface energies.

This simple 1d-example is generic in the sense that the 2 or 3d settings reproduce the same features as demonstrated in (Bouchitté et al., 1995), Section 3. An anti-plane shear result is immediately deduced from the results in that work. If the bulk energy W is isotropic (a function of the norm only), with p-growth ($p > 1$), and satisfies $W(0) = 0$, then, in the notation of Subsection 2.6, the lower semi-continuous envelope of

$$\mathcal{F}(\varphi) := \int_\Omega W(\nabla\varphi)\,dx + \int_{S(\varphi)} \kappa\big(\big|[\varphi]\big|\big)d\mathcal{H}^1$$

is given (see (2.30)) by

$$\mathcal{F}^*(\varphi) = \int_\Omega \hat{W}(\nabla\varphi)\,dx + \int_{S(\varphi)} \kappa\left(\big|[\varphi]\big|\right)d\mathcal{H}^1 + \sigma_c|C(\varphi)|, \qquad (4.19)$$

where $\hat{W}(F)$ is the inf-convolution of W with the linear mapping $F \mapsto \sigma_c F$, that is

$$\hat{W}(F) := \inf_{G,H}\{W(G) + \sigma_c H : G + H = F\}. \qquad (4.20)$$

By lower semi-continuous envelope, or relaxed energy, we mean – see e.g. the Appendix – the greatest function below $\mathcal{F}$ which is lower semi-continuous for the weak-* convergence in BV, i.e.,

$$\mathcal{F}^*(\varphi) = \inf_{\{\varphi_n\}} \liminf_n \{\ \mathcal{F}(\varphi_n) : \|\varphi_n\|_{BV(\Omega)} \leq \mathcal{C},$$

$$\varphi_n \to \varphi \text{ strongly in } L^1(\Omega)\}\,.$$

(The BV-norm of a BV-function φ is given by $\|\varphi\|_{L^1(\Omega)} + \int_\Omega |\nabla\varphi|\,dx + \int_{S(\varphi)} |[\varphi]|d\mathcal{H}^1 + |C(\varphi)|.$)

Note that it is immediate from the definition of the relaxed energy that, if $\{\varphi_n\}$ is such that

φ^n is a sequence of quasi-minimizers, i.e., $\mathcal{F}(\varphi^n) \leq \inf_\varphi \mathcal{F}(\varphi) + O(\frac{1}{n})$;

$\varphi^n \to \varphi$ strongly in $L^1(\Omega)$ with $\|\varphi^n\|_{BV(\Omega)} \leq \mathcal{C}$,

then φ is a minimizer for $\mathcal{F}^*$.

In view of the above, the relaxed bulk energy always grows *linearly* at infinity! Mechanicians are used to linearly growing energies. They come about in plasticity because the energy – as a function of the deformation – is the inf-convolution of the support function of the convex set of admissible stresses – the convex conjugate of the indicatrix function of that set, a linear function – with the elastic energy; see (Suquet, 1982). In this light, σ_c is a yield stress and relaxation induces bounded stresses, an essential component of plasticity.

We postpone until Section 7 a more detailed discussion of the exact relaxation in the cohesive setting (we have conveniently forgotten so far the presence of boundary conditions).

5. Irreversibility

Following the pattern adopted in the previous section, we address the case of a Griffith surface energy in a first subsection, then investigate its cohesive analogue in a second subsection.

5.1. Irreversibility – The Griffith case – Well-posedness of the variational evolution

In a Griffith setting, irreversibility is a simple notion: the crack can only extend with time,

$$\Gamma(t) \supset \Gamma(s), s < t.$$

With that notion in mind, we now discuss the variational evolution in a global minimality setting, noting that existence in such a context will automatically provide existence of that evolution for any kind of local minimality criterion. Once again, the argument put forth at the start of Paragraph 4.1.1 prohibits a wide range of force loads. We thus assume throughout this subsection that the only load is a displacement $g(t)$ defined on $\partial_d\Omega$, or rather, as we saw earlier in Subsection 2.5, on $\mathbb{R}^2$ while $f_s \equiv 0$ on $\partial_s\Omega \setminus \partial_d\Omega$.

5.1.1. *Irreversibility – The Griffith case – Discrete evolution*

As mentioned in the Introduction, the basic tool is also the natural computational tool: time discretization over the interval $[0, T]$. We thus consider

$$t_0 = 0 < t_1^n < \ldots\ldots < t_{k(n)}^n = T \text{ with } k(n) \overset{n}{\nearrow} \infty, \ \Delta_n := t_{i+1}^n - t_i^n \overset{n}{\searrow} 0.$$

Time-stepping the strong or weak minimality condition (Ugm), we obtain

(Sde) The strong discrete evolution: Find $(\Gamma_{i+1}^n, \varphi_{i+1}^n)$ a minimizer for

$$\min_{\varphi, \Gamma} \{ \int_{\Omega \setminus \Gamma} W(\nabla\varphi) \, dx + k\mathcal{H}^1(\Gamma \setminus \partial_s\Omega) : \varphi = g(t_{i+1}^n)$$
$$\text{on } \partial_d\Omega \setminus \Gamma; \ \Gamma \supset \Gamma_i^n \};$$

resp.

(Wde) The weak discrete evolution: Find φ_{i+1}^n a minimizer for

$$\min_{\varphi} \{ \int_{\Omega} W(\nabla\varphi) \, dx + k\mathcal{H}^1(S(\varphi) \setminus (\Gamma_i^n \cup \partial_s\Omega)) : \varphi = g(t_{i+1}^n)$$
$$\text{on } \partial_d\Omega \setminus S(\varphi) \};$$

then, $\Gamma_{i+1}^n = \Gamma_i^n \cup (S(\varphi_{i+1}^n) \setminus \partial_s\Omega)$.

$\varphi_n = 0$

Γ_n

$\varphi_n = 1$

Figure 5.1. The Neumann sieve

The balance (**Eb**) seems to have been forgotten all together in the discrete evolution, yet it will reappear in the time-continuous limit of those evolutions.

The first mathematical issue to confront is the existence of a solution to those discrete evolutions. As we mentioned before in Subsection 2.5, we cannot expect, even in 2d, a direct existence proof for the *strong discrete evolution* without imposing further restrictions on the class of admissible cracks. This is easily understood through the Neumann sieve example (Murat, 1985).

A Neumann sieve situation occurs when boundaries close up at a critical speed that creates channels of non–zero capacity in the domain. For example, consider $\Omega = (-1,1)^2$ and assume, in a linear anti-plane shear setting, that the crack Γ_n is given as $\{0\} \times [-1,1] \setminus \left(\bigcup_{p=-n,...,n} \left(\frac{p}{n} - e^{-n}, \frac{p}{n} + e^{-n} \right) \right)$ with

$$\varphi_n = \begin{cases} 0 \\ 1 \end{cases}, \text{ on } \begin{array}{l} \{x_1 = -1\} \\ \{x_1 = 1\}. \end{array}$$

Then φ_n satisfies

$$-\Delta \varphi_n = 0 \text{ on } \Omega_n := (-1,1)^2 \setminus \Gamma_n,$$

with $\partial \varphi_n / \partial \nu = 0$ on all boundaries of $\Omega \setminus \Gamma_n$, except $\{x_1 = \mp 1\}$. According to the results in (Murat, 1985) $\varphi_n \to \varphi$ strongly in $L^2(\Omega)$, with $\Omega = [(-1,0) \cup (0,1)] \times (-1,1)$ and φ is the solution, for some $\mu \neq 0$ of

$$-\Delta \varphi = 0 \text{ on } \Omega,$$

with

$$\begin{cases} \dfrac{\partial\varphi}{\partial x_2} = 0 & \text{on } \partial\Omega \cap \{x_2 = \pm 1\} \\[2mm] \varphi = 0, \text{ resp. } 1 & \text{on } \{-1\} \times (-1,1), \text{ resp. } \{1\} \times (-1,1) \\[2mm] \dfrac{\partial\varphi}{\partial x_1} = \mu[\varphi] & \text{on } \{0\} \times (-1,1). \end{cases}$$

Hence φ_n does not converge to the solution

$$\hat{\varphi} = 0 \ \text{ on } (-1,0) \times (-1,1); \ 1 \ \text{ on } (0,1) \times (-1,1)$$

of the Neumann problem on $\Omega\backslash\Gamma$, with $\Gamma = \{0\} \times (-1,1)$.

The Neumann sieve must thus be prevented so as to ensure the very existence of a pair-solution to the strong discrete evolution at each time step. A possible exit strategy consists in "prohibiting" disconnected cracks. A result in (Chambolle and Doveri, 1997) – see also (Bucur and Varchon, 2000) – states that, if Ω is a Lipschitz two dimensional domain and $\{\Gamma_n\}$ is a sequence of compact connected sets with $\mathcal{H}^1(\Gamma_n) \leq \mathcal{C}$ and such that it converges – for the Hausdorff metric – to Γ, the solution to a Neumann problem of the form

$$\begin{cases} -\Delta\varphi_n + \varphi_n = g & \text{in } \Omega\backslash\Gamma^n \\[2mm] \dfrac{\partial\varphi_n}{\partial\nu} = 0 & \text{on } \partial[\Omega\backslash\Gamma^n], \end{cases}$$

is such that $\varphi_n, \nabla\varphi_n \xrightarrow{n} \varphi, \nabla\varphi$, strongly in $L^2(\Omega)$, with φ the solution to

$$\begin{cases} -\Delta\varphi + \varphi = g & \text{in } \Omega\backslash\Gamma \\[2mm] \dfrac{\partial\varphi}{\partial\nu} = 0 & \text{on } \partial[\Omega\backslash\Gamma], \end{cases}$$

An adaptation of that result in (Dal Maso and Toader, 2002) proves the existence of a minimizer to the strong discrete evolution at each time step under the restriction that the cracks have an a priori set number of connected components. In turn, an analogous result is proved in (Chambolle, 2003) for plane elasticity.

Note that the connectedness restriction can be weakened to include cracks with an a priori set number of connected components (Dal Maso and Toader, 2002).

The discrete weak evolution behaves better as far as existence is concerned. Indeed, existence is a direct consequence of Ambrosio's compactness and lower semi-continuity results (2.24), (4.11) (see also Theorem D in the Appendix), or at least of a slight modification which consists in replacing $\mathcal{H}^1$ by $\mathcal{H}^1 \lfloor (\Gamma_i^n \cup \partial_s\Omega)^c$ in (2.24). To be precise, existence is established in

- The anti-plane shear case: φ is scalar-valued and W is convex with p-growth, $p > 1$;

- The case of non-linear elasticity: φ is vector-valued and W is quasi-convex with p-growth, $p > 1$. We refer the reader to the abundant literature on quasi-convexity (see e.g. (Ball and Murat, 1984)) and also to the Appendix for details on that notion; for our purpose, it suffices to remark that quasi-convexity, plus growth implies sequential weak lower semi-continuity on the Sobolev space $W^{1,p}(\Omega; \mathbb{R}^2)$ (Ball and Murat, 1984), but also, see (Ambrosio, 1994), for sequences $\{\varphi_n\}$ in $SBV(\Omega; \mathbb{R}^2)$ with

$$\begin{cases} \varphi_n \xrightarrow{L^1} \varphi \in SBV(\Omega; \mathbb{R}^2) \\ \mathcal{H}^1(S(\varphi_n)) \leq \mathcal{C} \end{cases}$$

It should be noted that the growth assumption prevents the energy density $W(F)$ from blowing up as $\det F \searrow 0$. But the latter is a desirable feature in hyperelasticity, at least according to popular wisdom. Once again, as in the comments following (2.26) in Subsection 2.5, we remain deliberately vague in this setting because of the subtle issues raised by the necessity of securing a supremum bound on the minimizing sequences, so as to apply Ambrosio's results.

Existence will not however be achieved in the setting of linearized elasticity which thus seems confined, for the time being, to the strong formulation.

Consider any setting for which the discrete evolution is meaningful. Then, for a given n (a given time step), we define the piecewise in time fields

$$\begin{cases} \varphi^n(t) := \varphi_i^n \\ \Gamma^n(t) := \Gamma_i^n \quad \text{on } [t_i^n, t_{i+1}^n), \text{ and, for } i = -1, \Gamma_{-1}^n := \Gamma_0. \\ g^n(t) = g(t_i^n) \end{cases}$$

Note that irreversibility is guaranteed at the discrete level because of the definition of Γ_i^n in terms of its predecessors. In other words,

$$\Gamma^n(t) \nearrow \text{ with } t.$$

Summing up, we have constructed, for each time $t \in [0, T]$, a pair $(\Gamma^n(t), \varphi^n(t))$ such that

(Sde) The strong discrete evolution: $(\Gamma^n(t), \varphi^n(t))$ is a minimizer for

$$\min_{\varphi, \Gamma} \Big\{ \int_{\Omega \backslash \Gamma} W(\nabla \varphi) \, dx + k \mathcal{H}^1(\Gamma \backslash \partial_s \Omega) :$$
$$\varphi = g^n(t) \text{ on } \mathbb{R}^2 \backslash \overline{\Omega}; \; S(\varphi) \subset \Gamma; \; \Gamma \supset \Gamma^n(t - \Delta_n) \Big\};$$

resp.

(Wde) The weak discrete evolution: $\varphi^n(t)$ is a minimizer for

$$\min_{\varphi} \Big\{ \int_{\Omega} W(\nabla \varphi) \, dx + k \mathcal{H}^1(S(\varphi) \backslash (\Gamma^n(t - \Delta_n) \cup \partial_s \Omega)) :$$
$$\varphi = g^n(t) \text{ on } \mathbb{R}^2 \backslash \overline{\Omega} \Big\}$$

and then $\Gamma^n(t) = \Gamma^n(t - \Delta_n) \cup S(\varphi^n(t))$.

Note that the time-discrete cracks in the previous formulation live on all of $\overline{\Omega}$ and not only on $\overline{\Omega} \backslash \partial_s \Omega$, but that there is no energy associated with the part of the crack that would live on $\partial_s \Omega$, as it should be. Also, here again, the functional dependence of $\varphi^n(t)$ is eschewed because it heavily depends upon the scalar/vectorial nature of the specific problem, as well as on the coercivity and growth properties of the bulk energy density W.

Note also that, at time $t = 0$, generically, it is not true that $\Gamma_0^n \equiv \Gamma_0$, but merely that $\Gamma_0^n \supset \Gamma_0$. There is an increase in the initial condition. Also, Γ_0^n is independent of n.

The goal is to pass to the limit in n and hope that the limit fields will be solutions to the strong/weak variational evolutions. As will be seen below, this is not a straightforward proposition.

5.1.2. *Irreversibility – The Griffith case – Global minimality in the limit*

A usual first step in a limit process is to obtain n-independent a priori estimates on the fields. This will be obtained here upon testing the strong/weak discrete evolutions **(Sde)**, **(Wde)** at each time by appropriate test fields. The two choice test fields are $(\Gamma^n(t), g^n(t))$ in the strong formulation, resp. $g^n(t)$ in the weak formulation, and $(\Gamma^n(t - \Delta_n), \varphi^n(t - \Delta_n) + g^n(t) - g^n(t - \Delta_n))$ in the strong formulation, resp. $\varphi^n(t - \Delta_n) + g^n(t) - g^n(t - \Delta_n)$ in the weak formulation (the addition of the terms involving g^n are so that the test deformations satisfy the boundary conditions at time t).

Then, provided we impose decent regularity on g, namely

$$g \in W^{1,1}(0, T; W^{1,p}(\Omega; \mathbb{R}^2)) \cap L^{\infty}((0, T) \times \Omega; \mathbb{R}^2)), \qquad (5.1)$$

for an energy with $p > 1$-growth, we obtain the following a priori bounds:

$$\begin{cases} \|\nabla\varphi^n(t)\|_{L^p(\Omega(;\mathbb{R}^2))} \leq \mathcal{C} \\ \mathcal{H}^1(S(\varphi^n(t)) \leq \mathcal{C} \text{ (weak formulation)}, \end{cases} \tag{5.2}$$

and

$$\mathcal{H}^1(\Gamma^n(t)) \leq \mathcal{C}, \tag{5.3}$$

together with the following upper bound on the total energy

$$\begin{aligned} E^n(t) &:= \int_\Omega W(\nabla\varphi^n(t))\, dx + k\mathcal{H}^1(\Gamma^n(t)\backslash\partial_s\Omega) \\ &\leq E^n(0) + \int_0^{\tau^n(t)}\int_\Omega \frac{\partial W}{\partial F}(\nabla\varphi^n(s))\cdot\nabla\dot{g}(s)\, dx\, ds + R_n, \end{aligned} \tag{5.4}$$

where $\tau^n(t) := \sup\{t_i^n \leq t\}$ and $R_n \to 0$. Note that the derivation of (5.4) actually requires a bit of care; see (Dal Maso et al., 2005), Section 6.

Can we pass to the n–limit in the minimality statements (Sde),(Wde) under the above convergences? And if so, is the result the expected variational evolution, or is the enterprise doomed by the specter of the Neumann sieve as more and more crack components accumulate at a given time when $n \nearrow$? For the strong formulation and under the connectedness restriction, the strong variational evolution is indeed obtained in the limit, as shown in (Dal Maso and Toader, 2002). We refer the interested reader to that reference and focus, from now onward in this paragraph, on the weak formulation.

To this effect, we first remark that the Neumann sieve phenomenon is merely a specter because the circumstances that presided over its appearance in Paragraph 5.1.1 were somewhat fallacious, for they failed to account for the role played by the surface energy. Indeed, assume that n is large enough; the pair (φ_n, Γ_n) considered in that example cannot be a joint minimizer of

$$\frac{1}{2}\int_{\Omega\backslash\Gamma} |\nabla\varphi|^2\, dx + \mathcal{H}^1(\Gamma), \ \Gamma \supset \Gamma_n$$

with the same boundary conditions. By lower semi-continuity,

$$\liminf_n \left\{\frac{1}{2}\int_{\Omega\backslash\Gamma} |\nabla\varphi_n|^2\, dx + \mathcal{H}^1(\Gamma_n)\right\} \geq \frac{1}{2}\int_{\Omega\backslash\Gamma} |\nabla\varphi|^2\, dx + 1,$$

with φ, the solution to the Neumann sieve. Now, φ has non zero bulk energy $\frac{1}{2}\int_{\Omega\backslash\Gamma} |\nabla\varphi|^2\, dx$, say $\mathcal{C} > 0$, so that, for n large enough,

$$\frac{1}{2}\int_{\Omega\backslash\Gamma} |\nabla\varphi_n|^2\, dx + \mathcal{H}^1(\Gamma_n) \geq 1 + \frac{\mathcal{C}}{2}.$$

But the energy associated with the pair $(\{0\} \times [-1, 1], \hat{\varphi})$ is exactly 1, a strictly smaller value, while $\{0\} \times [-1, 1] \supset \Gamma_n$. For n large enough, closing the holes of the sieve and taking the crack to be $\{0\} \times [-1, 1]$ is the energetically sound choice. This observation made Larsen and one of us (G.F.) hopeful for a derivation of the global minimality condition (Ugm) in the weak variational evolution from (Wde) under refinement of the time step.

That this is by no means a trivial endeavor can be illustrated as follows. We note first that, since $\mathcal{H}^1(B \backslash A) \geq \mathcal{H}^1(B) - \mathcal{H}^1(A)$, (Wde) implies in particular that $\varphi^n(t)$ is *a minimizer for its own jump set*, that is

$$\frac{1}{2} \int_\Omega W(\nabla \varphi^n(t)) \, dx \leq \frac{1}{2} \int_\Omega W(\nabla \varphi) \, dx + \mathcal{H}^1(S(\varphi) \backslash (S(\varphi^n(t)) \cup \partial \Omega_s)).$$
(5.5)

If (Ugm) is to be obtained in the limit, then $\varphi(t)$ should also in particular be a minimizer for its own jump set. In view of (5.2) and of the already quoted lower semi-continuity result of (Ambrosio, 1994), the left hand side of (5.5) is well behaved and the result would follow easily, provided that

$$\limsup_n \mathcal{H}^1(S(\varphi) \backslash S(\varphi^n(t))) \leq \mathcal{H}^1(S(\varphi) \backslash S(\varphi(t))).$$

Consider however φ such that $S(\varphi) \subset S(\varphi(t))$, while the jump set of $\varphi^n(t)$ does not intersect that of $\varphi(t)$ (which would surely happen if $S(\varphi^n(t)) \subset K_n$, with $K_n \cap K = \emptyset$ and the Hausdorff distance from K_n to K goes to 0). Then $\mathcal{H}^1(S(\varphi))$ must be 0!

The stability of the "own jump set minimality condition" cannot be established without a modification of the test fields φ. This is the essence of the jump transfer theorem (Francfort and Larsen, 2003), Section 2. We now quote it without proof in its simplest version, emphasizing its decisive role in establishing existence of the weak variational evolution (see Theorem 5.5 below).

THEOREM 5.1. *Let* $\varphi^n, \varphi \in SBV(\Omega)$ *with* $\mathcal{H}^1(S(\varphi)) < \infty$, *be such that*

- $|\nabla \varphi^n|$ *weakly converges in* $L^1(\Omega)$; *and*

- $\varphi^n \rightarrow \varphi$ *in* $L^1(\Omega)$.

Then, for every $\zeta \in SBV(\Omega)$ *with* $\nabla \zeta \in L^p(\Omega)$, $1 \leq p < \infty$, *and* $\mathcal{H}^1(S(\zeta)) < \infty$, *there exists a sequence* $\{\zeta^n\} \subset SBV(\Omega)$ *with* $\nabla \zeta^n \in L^p(\Omega)$, *such that*

- $\zeta^n \rightarrow \zeta$ *strongly in* $L^1(\Omega)$;

- $\nabla\zeta^n \to \nabla\zeta$ *strongly in $L^p(\Omega)$; and*

- $\limsup_n \mathcal{H}^1 \lfloor A\left(S(\zeta^n)\backslash S(\varphi^n)\right) \leq \mathcal{H}^1 \lfloor A\left(S(\zeta)\backslash S(\varphi)\right)$, *for any Borel set A.*

We fix a time t and recall (5.2). Ambrosio's compactness result permits one to assert the existence of a t-dependent subsequence $\{\varphi^{n_t}(t)\}$ of $\{\varphi^n(t)\}$ and of $\varphi(t)$ such that the assumptions of Theorem 5.1 – or rather of a corollary of Theorem 5.1 which takes into account the boundary conditions on the test fields at t, namely $\varphi^{n_t}(t) = g^{n_t}(t)$ on $\mathbb{R}^2\backslash\Omega$ – are met. The conclusion of that theorem then allows for a corresponding sequence $\{\zeta^{n_t}\}$ that is an admissible test in (Wde), so that

$$\int_\Omega W(\nabla\varphi^{n_t}(t))\,dx \leq \int_\Omega W(\nabla\zeta^{n_t})\,dx + \mathcal{H}^1\left(S(\zeta^{n_t})\backslash(S(\varphi^{n_t}(t)) \cup \partial_s\Omega)\right),$$

and then, from the convergences obtained in the theorem, together with the assumed p-growth of the energy, we pass to the limit in n_t and obtain that the limit $\varphi(t)$ is a minimizer for its own jump set, that is

$$\int_\Omega W(\nabla\varphi(t))\,dx \leq \int_\Omega W(\nabla\zeta)\,dx + \mathcal{H}^1\left(S(\zeta)\backslash(S(\varphi(t)) \cup \partial_s\Omega)\right).$$

We are inching ever closer to the global minimality statement (Ugm) in the weak variational evolution, but are not quite there yet, because we would like to remove not only $S(\varphi(t)) \cup \partial_s\Omega$ but $\Gamma(t) \cup \partial_s\Omega$ in the minimality statement above. To do this, we need to define the limit crack $\Gamma(t)$. There are various setting-dependent paths to a meaningful definition of the limit crack. An encompassing view of that issue is provided by the notion of σ_p-convergence introduced in (Dal Maso et al., 2005), Section 4, a kind of set convergence for lower dimensional sets.

DEFINITION 5.2. Γ^n σ_p-*converges to Γ if $\mathcal{H}^1(\Gamma^n)$ is bounded uniformly with respect to n, and*

(1) whenever $\varphi^j, \varphi \in SBV(\mathbb{R}^2)$ are such that

$$\begin{cases} \varphi^j \xrightarrow{weak-*} \varphi, \ in \ L^\infty(\mathbb{R}) \\[2mm] \nabla\varphi^j \xrightarrow{weak} \nabla\varphi, \ in \ L^p(\mathbb{R}^2) \\[2mm] S(\varphi^j) \subset \Gamma^{n_j} \end{cases}$$

for some sequence $n_j \nearrow \infty$, then $S(\varphi) \subset \Gamma$;

(2) there exist a function $\varphi \in SBV^p(\mathbb{R}^2)$ with $S(\varphi) = \Gamma$ and a sequence φ^n with the properties of item (1).

REMARK 5.3. Note that it is immediate from item (2) in the above definition and from (2.24) that $\mathcal{H}^1(\Gamma) \leq \liminf_n \mathcal{H}^1(\Gamma_n)$.

Then, the following compactness result proved in (Dal Maso et al., 2005), Section 4.2, holds true:

THEOREM 5.4. *Let $\Gamma^n(t)$ be a sequence of increasing sets defined on $[0,T]$ and contained in a bounded set B. Assume that $\mathcal{H}^1(\Gamma^n(t))$ is bounded uniformly with respect to n and t. Then there exist a subsequence $\Gamma^{n_j}(t)$ and a $\Gamma(t)$ defined on $[0,T]$ such that*

$$\Gamma^{n_j}(t) \quad \sigma_p\text{-converges to } \Gamma(t), \ \forall t \in [0,T].$$

The estimate (5.3) permits one to apply the theorem above and thus to define a meaningful crack $\Gamma(t)$ such that, for a subsequence still labeled $\Gamma^n(t)$, $\Gamma^n(t)$ σ_p–converges to $\Gamma(t)$, hence also $\Gamma^{n_t}(t)$. Because of item (2) in Definition 5.2, we can construct φ with $S(\varphi) = \Gamma(t)$ and φ^{n_t} satisfying the assumptions of Theorem 5.1 with $S(\varphi^{n_t}) \subset \Gamma^{n_t}(t)$. But (Wde) implies in particular that

$$\int_\Omega W(\nabla\varphi^{n_t}(t)) \, dx \leq \int_\Omega W(\nabla\zeta) \, dx + k\mathcal{H}^1(S(\zeta)\backslash(\Gamma^{n_t}(t) \cup \partial_s\Omega))$$

$$\leq \int_\Omega W(\nabla\zeta) \, dx + k\mathcal{H}^1(S(\zeta)\backslash(S(\varphi^{n_t}) \cup \partial_s\Omega)).$$

and the jump transfer Theorem 5.1 delivers the required minimality property (Ugm).

Having obtained global minimality, we are still faced with the question of the validity of the energy conservation statement (Eb). This is the object of the next paragraph.

5.1.3. *Irreversibility – The Griffith case – Energy balance in the limit*
Inequality (5.4) derived at the onset of Paragraph 5.1.2 hints at the possibility of an energy inequality. To obtain such an inequality in the limit, it suffices, in view of Remark 5.3 to ensure that, as $n_t \nearrow \infty$,

$$\int_\Omega W(\nabla\varphi^{n_t}(t)) \, dx \rightarrow \int_\Omega W(\nabla\varphi(t)) \, dx \qquad (5.6)$$

and that

$$\limsup_{n_t} \int_0^{\tau^n(t)} \int_\Omega \frac{\partial W}{\partial F}(\nabla\varphi^{n_t}(s)) \cdot \nabla\dot{g}(s) dx \, ds \leq$$

$$\int_0^t \int_\Omega \frac{\partial W}{\partial F}(\nabla\varphi(s)) \cdot \nabla\dot{g}(s) \, dx \, ds.$$

$$(5.7)$$

Equality (5.6) is nearly immediate; one inequality holds true by lower semi-continuity as seen several times before. The other is obtained upon applying the jump transfer Theorem 5.1 to $\varphi(t)$ itself and inserting the resulting test sequence in (5.5). This yields the other inequality, namely

$$\limsup_{n_t} \int_\Omega W(\nabla\varphi^{n_t}(t)) \, dx \leq \int_\Omega W(\nabla\varphi(t)) \, dx.$$

The derivation of (5.7) is more involved in the non-quadratic case. Indeed, it amounts, modulo application of Fatou's lemma for the time integral, to showing that the stresses $\partial W/\partial F(\nabla\varphi^{n_t}(t))$ converge weakly to the limit stress $\partial W/\partial F(\nabla\varphi(t))$. Although a surprising result, this is indeed the case in view of the convergences announced for $\varphi^{n_t}(t)$ to $\varphi(t)$ and of (5.6); we omit the proof and refer the interested reader to (Dal Maso et al., 2005), Section 4.3. The following energy inequality is established:

$$
\begin{aligned}
E(t) &:= \int_\Omega W(\nabla\varphi(t)) \, dx + k\mathcal{H}^1(\Gamma(t)\backslash\partial_s\Omega) \\
&\leq E(0) + \int_0^t \int_\Omega \frac{\partial W}{\partial F}(\nabla\varphi(s)) \cdot \nabla\dot{g}(s) \, dx \, ds,
\end{aligned}
\tag{5.8}
$$

The other energy inequality is a byproduct of the minimality statement (Ugm). Simply test global minimality at time s by $\varphi(t) + g(s) - g(t)$, $t > s$. Then, since $S(\varphi(t)) \subset \Gamma(t)$,

$$
\begin{aligned}
\int_\Omega W(\nabla\varphi(s)) \, dx &\leq \int_\Omega W(\nabla\varphi(t) + g(s) - g(t)) \, dx \\
&\quad + \mathcal{H}^1(S(\varphi(t))\backslash(\Gamma(s) \cup \partial_s\Omega)) \\
&\leq \int_\Omega W(\nabla\varphi(t) + g(s) - g(t)) \, dx \\
&\quad + \mathcal{H}^1(\Gamma(t)\backslash(\Gamma(s) \cup \partial_s\Omega)) \\
&= \int_\Omega W(\nabla\varphi(t)) \, dx + \mathcal{H}^1(\Gamma(t)\backslash(\Gamma(s) \cup \partial_s\Omega)) \\
&\quad - \int_\Omega \frac{\partial W}{\partial F}\left(\nabla\varphi(t) + \rho(s,t) \int_s^t \nabla\dot{g}(\tau) \, d\tau\right) \cdot \\
&\qquad \int_s^t \nabla\dot{g}(\tau) \, d\tau \, dx,
\end{aligned}
$$

for some $\rho(s,t) \in [0,1]$. Hence

$$E(t) - E(s) \geq \int_\Omega \frac{\partial W}{\partial F}\left(\nabla\varphi(t) + \rho(s,t) \int_s^t \nabla\dot{g}(\tau) \, d\tau\right) \cdot \int_s^t \nabla\dot{g}(\tau) \, d\tau \, dx.$$

We then choose a partition $0 < s_1^n < .... < s_{k(n)}^n = t$ of $[0, t]$, with $\Delta_n' := s_{i+1}^n - s_i^n \searrow 0$; summing the contributions, we get

$$E(t) - E(0) \geq \sum_{i=0}^{k(n)} \int_\Omega \frac{\partial W}{\partial F} \left(\nabla\varphi(s_{i+1}^n) + \rho(s_i^n, s_{i+1}^n) \int_{s_i^n}^{s_{i+1}^n} \nabla\dot{g}(\tau) \, d\tau \right)$$

$$\cdot \int_{s_i^n}^{s_{i+1}^n} \nabla\dot{g}(\tau) d\tau dx.$$

A uniform continuity type result – already implicitly used in the derivation of (5.4) – permits us to drop the term depending on $\rho(s_i^n, s_{i+1}^n)$ in the previous inequality in the limit $\Delta_n' \searrow 0$; see (Dal Maso et al., 2005), Section 4.3. Thus

$$E(t) - E(0) \geq \limsup_n \left\{ \sum_{i=0}^{k(n)} \int_{s_i^n}^{s_{i+1}^n} \int_\Omega \frac{\partial W}{\partial F} \left(\nabla\varphi(s_{i+1}^n) \right) \cdot \nabla\dot{g}(\tau) \, dx \, d\tau \right\}.$$

The expression on the right hand-side of the previous inequality looks very much like a Riemann sum. A not so well-known result in integration asserts that Riemann sums of a Lebesgue integrable function do converge to the integral of that function, but only for carefully chosen partitions (Hahn, 1914). Since we enjoy complete liberty in our choice of the partition $\{s_j^n\}$ of $[0, t]$, we conclude that

$$E(t) - E(0) \geq \int_0^t \int_\Omega \frac{\partial W}{\partial F} \left(\nabla\varphi(s) \right) \cdot \nabla\dot{g}(\tau) \, dx \, d\tau,$$

which, together with (5.8), provides the required equality (Eb).

5.1.4. *Irreversibility – The Griffith case – The time-continuous evolution*

Here, the results obtained in the previous paragraphs are coalesced into an existence statement to the weak variational evolution. The result is expressed in a 2d setting, but it applies equally in a 3d setting, upon replacing $\mathcal{H}^1$ by $\mathcal{H}^2$. We also recall similar existence results obtained in (Dal Maso and Toader, 2002), (Chambolle, 2003) in the 2d connected case.

In what follows, the energy density W is a nonnegative convex – in the anti-plane shear setting – or quasiconvex – in the plane setting – C^1 function on $\mathbb{R}^2$ with

$$(1/\mathcal{C})|F|^p - \mathcal{C} \leq W(F) \leq \mathcal{C}|F|^p + \mathcal{C}, \quad \forall F, \ 1 < p < \infty.$$

Note that the assumptions on W immediately imply that (see e.g. (Dacorogna, 1989))

$$|DW(F)| \leq \mathcal{C}(1 + |F|^{p-1}).$$

The domain Ω under consideration is assumed throughout to be Lipschitz and bounded, and the function g, which appears in the boundary condition on $\partial\Omega_d$, is assumed to be defined on all of $\mathbb{R}^2$; actually, each of its components is taken to be in $W_{loc}^{1,1}([0, \infty); W^{1,p}(\mathbb{R}^2))$.

The traction-free part $\partial_s\Omega$ of the boundary is assumed to be closed. Finally, the pre-existing crack Γ_0 is a closed set in Ω, with $\mathcal{H}^1(\Gamma_0) < \infty$.

THEOREM 5.5. $\exists \Gamma(t) \subset \overline{\Omega}$ and φ such that

(1) Each component of $\varphi(t) \in SBV(\mathbb{R}^2)$, with $\nabla\varphi$ p-integrable;

(2) $\Gamma(t) \supset \Gamma_0$ increases with t and $\mathcal{H}^1(\Gamma(t)) < +\infty$;

(3) $S(\varphi(t)) \subset \Gamma(t) \cup \partial_s\Omega$ and $\varphi(t) = g(t)$ a.e. on $\mathbb{R}^2\backslash\overline{\Omega}$;

(4) For every $t \geq 0$ the pair $(\varphi(t), \Gamma(t))$ minimizes

$$\int_\Omega W(\nabla\varphi)\, dx + \mathcal{H}^1(\Gamma\backslash\partial_s\Omega)$$

among all $\Gamma \supset \Gamma(t)$ and φ with components in $SBV(\mathbb{R}^2)$ s.t. $S(\varphi) \subset \Gamma$ and $\varphi = g(t)$ a.e. on $\mathbb{R}^2\backslash\overline{\Omega}$;

(5) The total energy

$$E(t) := \int_\Omega W(\nabla\varphi(t))\, dx + \mathcal{H}^1(\Gamma(t)\backslash\partial_s\Omega)$$

is absolutely continuous, $DW(\nabla\varphi)\cdot\nabla\dot{g} \in L_{loc}^1([0, \infty); L^1(\mathbb{R}^2))$, and

$$E(t) = E(0) + \int_0^t \int_\Omega DW(\nabla\varphi(s)) \cdot \nabla\dot{g}(s)\, dx\, ds.$$

As remarked before in Subsection 4.1, we did not wish to incorporate in this study body or surface loads because of the complex structure of the allowed class of soft devices introduced in (Dal Maso et al., 2005), Section 3. Also, in the vector-valued setting, it is assumed that somehow, the deformations are always capped in sup-norm by some set number. This is an a priori assumption which can be verified for certain classes of quasi-convex energies (Leonetti and Siepe, 2005). Note that there is no need for such an assumption in the anti-plane shear case, provided that the displacement load g is also bounded in sup-norm.

In 2d only and in the case where the cracks are a priori assumed to be connected – or to have a pre-set number of connected components – then the same existence result for the strong variational evolution is obtained in (Dal Maso and Toader, 2002) in the quadratic case and in (Chambolle, 2003) in the case of linearized elasticity. The statement is identical to that of Theorem 5.5 at the expense of replacing $\int_\Omega$ by $\int_{\Omega\backslash\Gamma}$, and considering φ's with components in $L^{1,2}(\Omega\backslash\Gamma) := \{f \in L^2_{loc}(\Omega\backslash\Gamma) : \nabla f \in L^2(\Omega\backslash\Gamma)\}$, resp. $\varphi \in LD(\Omega\backslash\Gamma) := \{\zeta \in L^2_{loc}(\Omega\backslash\Gamma; \mathbb{R}^2) : e(\zeta) \in L^2(\Omega\backslash\Gamma; \mathbb{R}^4)\}$, in the case of linear elasticity.

This existence result calls for comments. First and foremost, it is an existence result, not a uniqueness result. As in other non-convex problems in mechanics, uniqueness should not be expected: just think of the example of the elastic strip in Proposition 4.5 where the cracked section can be any vertical section of the sample.

Then the regularity of the field $\varphi(t)$, or lack thereof, is precious information. It indicates that time jumps could appear in the various fields. Indeed, still referring to that same example, we witness there a brutal decrease to 0 at time t_i of the bulk energy with a corresponding increase of the surface energy. This is precisely what (Eb) is about: a conspiracy of jumps that will remain undetected by the total energy.

Third, an implicit change of initial conditions may occur, since it might happen that $\Gamma(0)$ contains, but does not equal Γ_0. It is our belief, substantiated by the results of Proposition 4.3, that such a brutal event will not take place, but we have no proof at present.

Finally, the weak evolution might just turn out to be a strong evolution in disguise, as was the case for image segmentation in the light of the results of (De Giorgi et al., 1989), in which case there would be no need for the strong variational evolution. But wishing it so does not make it so, and the task at hand is forbidding.

REMARK 5.6. The unilateral global minimality condition (item 4. in Theorem 5.5) can actually be strengthened as follows:
For every $t \geq 0$ the pair $(\varphi(t), \Gamma(t))$ minimizes

$$\int_\Omega W(\nabla\varphi)\, dx + \mathcal{H}^1(\Gamma\backslash\partial_s\Omega)$$

among all $\Gamma \supset \cup_{s<t}\Gamma(s)$ and φ with components in $SBV(\mathbb{R}^2)$ s.t. $S(\varphi) \subset \Gamma$ and $\varphi = g(t)$ a.e. on $\mathbb{R}^2\backslash\overline{\Omega}$.

This states that the admissible test cracks do not have to contain the current crack, but only those up to, but not including the current time, a clearly stronger minimality condition. The two conditions are actually equivalent because, for $s < t$, unilateral global minimality implies in particular that

$$\int_\Omega W(\nabla\varphi(s))\, dx + \mathcal{H}^1(\Gamma(s)\backslash\partial_s\Omega) \leq \int_\Omega W(\nabla\varphi + \nabla g(s) - \nabla g(t))\, dx$$
$$+ \mathcal{H}^1(\Gamma\backslash\partial_s\Omega),$$

for any φ with components in $SBV(\mathbb{R}^2)$ s.t. $\varphi = g(t)$ a.e. on $\mathbb{R}^2\backslash\overline{\Omega}$, $S(\varphi) \subset \Gamma$ and any $\Gamma \supset \cup_{s<t}\Gamma(s)$. Let $s \nearrow t$ and use item 5. (the

continuity of the total energy) to pass to the limit in the left hand-side of the inequality above. The stronger minimality result is then obtained by dominated convergence (since W has p–growth).

To conclude this section, we refer the reader to the numerical example developed in Paragraph 8.3.3, which illustrates the issues that we have tackled so far – initiation and irreversibility – in the context of global minimality. The brutal onset of the cracking process evidenced in Figure D-b, page 102, agrees with the result obtained in Proposition 4.3 because the crack appears at a non-singular point, thus the initiation time must be positive and the onset brutal. Also note from Figure C, page 102, that the energy is conserved during the phases of brutal growth, as theoretically expected because the total energy should in particular be continuous in time.

5.2. IRREVERSIBILITY – THE COHESIVE CASE

In the Griffith case, irreversibility is a purely geometric issue: the crack at the current time is in essence the union of all discontinuities of the kinematic variable throughout its past. By contrast, in the cohesive case, the cohesive forces should somehow reflect the complete history of the deformation undergone by the material up to the present time. Cohesive forces are affected by the magnitude of those discontinuities, and not only by their mere presence. Thus, the choice of the right memory variable is crucial.

Our benchmark example throughout this Subsection is the square pre-cracked sample Ω in Figure 5.2. It is filled with an isotropic material with energy density W (endowed with the usual properties) and loaded in mode-I by a displacement load $f(t)$ as shown in Figure 5.2. The surface energy $\kappa(\lambda, \tau)$ is as in Paragraph 4.2.2, and it is assumed to be differentiable, with partial derivatives respectively denoted by $\partial \kappa / \partial \lambda, \partial \kappa / \partial \tau$. Symmetry implies that the crack – understood as the locus of the possible discontinuities of the kinematic field – will live in $\hat{\Gamma} := [0, L] \times \{0\}$ and that $[\varphi]$ is parallel to $\vec{e}_2$, so that we will identify $[\varphi]$ with its vertical component. Note that, if σ denotes the stress field $DW(\nabla \varphi(t))$, then $\sigma_{12} = 0$ for $x_2 = 0$.

Adopt as memory variable the *maximal opening*, that is

$$\psi(t, x) := \sup_{s \leq t} [\varphi(s, x)], \quad \text{on } \hat{\Gamma}. \tag{5.9}$$

The surface energy at t is $\int_{\hat{\Gamma}} \kappa(\psi(t), 0) \, d\mathcal{H}^1$. In the spirit of Paragraph 5.1.1, we now investigate an incremental evolution of the crack. As mentioned in Subsection 4.2, we do not have to invoke minimality

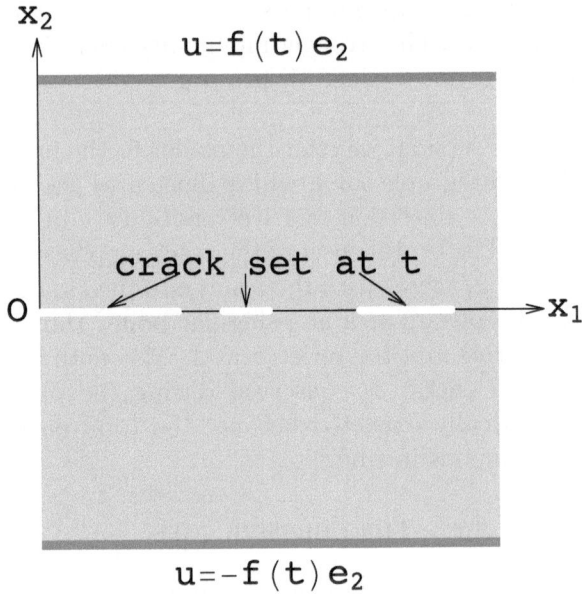

Figure 5.2. Sample loaded in mode I

here, and simply impose unilateral stationarity, a notion defined in Section 2. The problem becomes, with notation that should be familiar to the reader at this stage,

Find a unilateral stationary point φ_i for

$$\int_{\Omega \backslash \hat{\Gamma}} W(\nabla \varphi) \, dx + \int_{\hat{\Gamma}} \kappa(\max\{\psi_{i-1}, [\varphi]\}, 0) \, d\mathcal{H}^1.$$

The associated stress field σ_i must satisfy

$$\begin{cases} \operatorname{div} \sigma_i = 0 \text{ in } \Omega \backslash \hat{\Gamma} \\[2mm] \sigma_i \, e_1 = 0 \text{ on the part of } \partial\Omega \text{ with normal } \pm e_1. \end{cases} \tag{5.10}$$

We need to compute the cohesive forces. To this effect define

$$\begin{aligned} \hat{\Gamma}_i^+ &= \{x \in \hat{\Gamma} : [\varphi_i(x)] > \psi_{i-1}(x)\} \\ \hat{\Gamma}_i^- &= \{x \in \hat{\Gamma} : [\varphi_i(x)] < \psi_{i-1}(x)\} \\ \hat{\Gamma}_i^0 &= \{x \in \hat{\Gamma} : [\varphi_i(x)] = \psi_{i-1}(x)\}, \end{aligned}$$

and assume that both ψ_{i-1} and φ_i are smooth enough to lend meaning to the expressions below; this will be the case if e.g. W has $p > 2$-growth, in which case those quantities will be continuous on $\hat{\Gamma}$.

Because of the equilibrium equations (5.10), unilateral stationarity implies that, for all ζ's with $\zeta \in L^\infty(\Omega;\mathbb{R}^2)$ and $S(\zeta) \subset \hat{\Gamma}$,

$$\int_{\hat{\Gamma}} (\sigma_i)_{22}[\zeta_2]\,dx_1 \leq \int_{\hat{\Gamma}_i^+} \frac{\partial\kappa}{\partial\lambda}(\psi_i,0)[\zeta_2]\,dx_1 + \int_{\hat{\Gamma}_i^0} \frac{\partial\kappa}{\partial\lambda}(\psi_i,0)[\zeta_2]^+\,dx_1.$$

Then, by the arbitrariness of ζ,

$$(\sigma_i)_{22} = \frac{\partial\kappa}{\partial\lambda}(\psi_i,0) \text{ on } \hat{\Gamma}_i^+$$

$$(\sigma_i)_{22} = 0 \text{ on } \hat{\Gamma}_i^-$$

$$0 \leq (\sigma_i)_{22} \leq \frac{\partial\kappa}{\partial\lambda}(\psi_i,0) \text{ on } \hat{\Gamma}_i^0.$$

Consequently, irreversibility in the form of a maximal opening criterion cancels all cohesive forces as long as the current opening does not exceed that at all previous times.

REMARK 5.7. In (Dal Maso and Zanini, 2007), the existence of a time-continuous evolution in a cohesive setting with the maximal opening as memory variable is established under the assumption (Ugm) and provided that the crack site is constrained to live on a smooth manifold of co-dimension 1. The proof is based, once again, on a time-stepping process. This justifies, at least in the case of global minimality, the incremental framework adopted in this subsection.

We propose to investigate the response of the material during cyclic loading, that is when $f(t)$ has a seesaw-type time variation as in Figure 5.3. This is the litmus test for fatigue. During the first loading phase, the opening $[\varphi(x,t)]$ will monotonically increase with time throughout $\hat{\Gamma}$, as will be proved in Section 9, at least in a simplified context. Denote by $(\varphi_1,\sigma_1,\psi_1)$ the respective values of the kinematic field, the stress field and the maximal opening at the end of the first loading phase. As unloading occurs, intuition, corroborated by the results of Section 9, strongly militates for a decrease in maximal opening, so that the surface energy will remain unchanged during that phase. The second loading phase will merely result in a maximal opening ψ_1, so that the material response will be that experienced during the first cycle. And this ad nauseam, thus forbidding the onset of fatigue.

In the framework of maximal opening "summon[ing] up remembrance of things past, [we] sigh the lack of many a thing [we] sought.... and moan the expense of many a vanished sight"[9].

---

[9] Shakespeare – Sonnet XXX

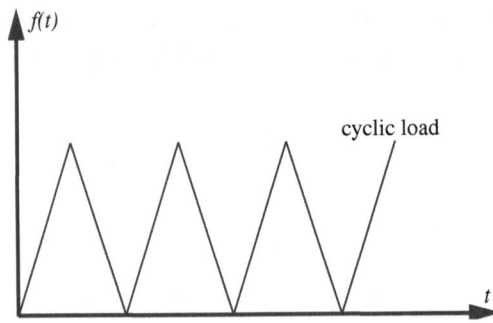

Figure 5.3. Seesaw-type cyclic load

Adopt then as memory variable the *cumulated opening*, that is, in lieu of (5.9),

$$\psi(x,t) = \int_0^t ([\dot{\varphi}(\overset{\circ}{x},s)])^+ \, ds \text{ on } \hat{\Gamma}, \tag{5.11}$$

where the dot denotes the time derivative and the $+$-sign stands for the positive part.

The surface energy at t is $\int_{\hat{\Gamma}} \kappa(\psi(t),0)d\mathcal{H}^1$. The incremental *cumulated* opening then becomes

$$\psi_i = \psi_{i-1} + ([\varphi_i] - [\varphi_{i-1}])^+,$$

and the total energy at time t_i may be written as

$$\int_{\Omega \setminus \hat{\Gamma}} W(\nabla\varphi) \, dx + \int_{\hat{\Gamma}} \kappa(\psi_{i-1} + ([\varphi] - [\varphi_{i-1}])^+, 0) \, d\mathcal{H}^1.$$

Unilateral stationarity requires once more to partition $\hat{\Gamma}$ into various parts that compare $[\varphi_i]$ to $[\varphi_{i-1}]$ and the following conditions are derived:

$$(\sigma_i)_{22} = \frac{\partial \kappa}{\partial \lambda}(\psi_i, 0) \text{ on } \hat{\Gamma}_i^+ = \{x \in \hat{\Gamma} : [\varphi_i](x) > [\varphi_{i-1}](x)\}$$

$$(\sigma_i)_{22} = 0 \text{ on } \hat{\Gamma}_i^- = \{x \in \hat{\Gamma} : [\varphi_i](x) < [\varphi_{i-1}](x)\}$$

$$0 \le (\sigma_i)_{22} \le \frac{\partial \kappa}{\partial \lambda}(\psi_i, 0) \text{ on } \hat{\Gamma}_i^0 = \{x \in \hat{\Gamma} : [\varphi_i](x) = [\varphi_{i-1}](x)\}.$$

At first glance, these conditions are similar to those obtained for a maximal opening criterion. However, the definition of the various domains has changed and a more careful examination shows that irreversibility only cancels the cohesive forces if the opening actually decreases. Each time the opening increases, the surface energy evolves and the cohesive forces are obtained as derivatives of that surface energy.

During cyclic loading (see Figure 5.3) the sample's behavior will be drastically altered. Indeed, during the first loading phase the opening grows and $\psi(t) = [\varphi(t)]$. The material response is indistinguishable from that previously obtained. The same holds true of the second part of the first cycle which corresponds to a phase of unloading. But, during the second loading phase, the opening increases again and the surface energy will evolve, in contrast to what takes place for the maximal opening criterion. Thus, with the cumulated opening as memory variable, we at least "stand a ghost of a chance"[10] with fatigue.

REMARK 5.8. The specific geometry and loading of the imaginary sample used in this subsection has allowed us to focus on the normal component of the displacement at the site of the possible discontinuities. Further, non-interpenetration was automatically enforced because the displacement load $f(t)$ forces the lips of the potential crack to open. In a more general setting, such would not be the case.

Non-interpenetration could be systematically imposed by only allowing non-negative normal jumps. The issue of the correct choice for a memory variable should be raised for the tangential jumps as well. If contemplating cumulated opening as the correct memory variable for sliding, then all slides, whatever their signs, should contribute, so that the positive part of the derivative of the jump in (5.11) should be replaced by the absolute value of that derivative.

Assuming only sliding occurs, then all test fields will be such that $\varphi \perp \nu$, with ν normal to $\hat{\Gamma}$ and the incremental cumulated sliding is

$$\gamma_i = \gamma_{i-1} + |\varphi_i - \varphi_{i-1}|,$$

while the surface energy for a test slide is $\int_{\hat{\Gamma}} \kappa(0, \gamma_{i-1} + |\varphi - \varphi_{i-1}|) \, d\mathcal{H}^1$. Unilateral stationarity then yields, strictly as before,

$$(\sigma_i)_{12} = \text{sign} \, (\varphi_i - \varphi_{i-1}) \frac{\partial \kappa}{\partial \tau}(0, \gamma_i) \text{ on } \hat{\Gamma}_i^{\pm} = \{x \in \hat{\Gamma} : \varphi_i(x) \neq \varphi_{i-1}(x)\}$$

$$|(\sigma_i)_{12}| \leq \frac{\partial \kappa}{\partial \tau}(0, \gamma_i) \text{ on } \hat{\Gamma}_i^0 = \{x \in \hat{\Gamma} : \varphi_i(x) = \varphi_{i-1}(x)\}.$$

The reader, gently prodded by the previous arguments, will now undoubtedly acquiesce to cumulative opening as the correct measure of irreversibility in a cohesive setting. She will consequently not object to the general setting for cohesive crack growth proposed below in the 2d case.

---

[10] Victor Young – composer, 1932

- **The weak cohesive variational evolution revisited** : Find, for every $t \in [0, T]$, $(\Gamma(t), \varphi(t))$ satisfying

(Ulm) $(\Gamma(t), \varphi(t))$ is a local minimizer (in a topology that remains to be specified) for

$$\mathcal{E}(t; \varphi, \Gamma) := \int_\Omega W(\nabla\varphi)dx - \mathcal{F}(t, \varphi) +$$

$$\int_{\Gamma(t)} \kappa(\psi(t) + [(\varphi - \varphi(t)) \cdot \nu]^+, \; \gamma(t) + |[(\varphi - \varphi(t)) \times \nu]|)d\mathcal{H}^1$$

among all $\overline{\Omega} \backslash \partial_s \Omega \supset \Gamma \supset \Gamma(t)$ and all $\varphi \equiv g(t)$ on $\mathbb{R}^2 \backslash \overline{\Omega}$ with $S(\varphi) \subset \Gamma$; or, resp.,

(Ugm) $(\Gamma(t), \varphi(t))$ is a global minimizer for $\mathcal{E}(t; \varphi, \Gamma)$ among all $\overline{\Omega} \backslash \partial_s \Omega \supset \Gamma \supset \Gamma(t)$ and all $\varphi \equiv g(t)$ on $\mathbb{R}^2 \backslash \overline{\Omega}$ with $S(\varphi) \subset \Gamma$;

(Eb) $\dfrac{dE}{dt}(t) = \displaystyle\int_\Omega \dfrac{\partial W}{\partial F}(\nabla\varphi(t)) \cdot \nabla\dot{g}(t) \, dx - \dot{\mathcal{F}}(t, \varphi(t)) - \mathcal{F}(t, \dot{g}(t))$

with

$$E(t) = \int_\Omega W(\nabla\varphi(t)) \, dx - \mathcal{F}(t, \varphi(t)) + \int_{\Gamma(t)} \kappa(\psi(t), \gamma(t))d\mathcal{H}^1.$$

Above, the undefined quantities are

$$\psi(t) := \sum_{\{t_i\} \text{ partitions of } [0,t]} [(\varphi(t_{i+1}) - \varphi(t_i)) \cdot \nu_{i+1}]^+,$$

$$\gamma(t) := \sum_{\{t_i\} \text{ partitions of } [0,t]} |[(\varphi(t_{i+1}) - \varphi(t_i)) \times \nu_{i+1}]|$$

(where ν_i is the normal (at a given point) to the jump set $S(\varphi(t_i))$).

REMARK 5.9. Once the crack path has been unconstrained, the issue of stationarity versus minimality pops up again. Our statement of the weak cohesive variational evolution adopts minimality. This is rather inconsequential, because, as explained several times before, our mathematical grasp of that kind of evolution is rudimentary at best. If initiation could be discussed with some rigor in Subsection 4.2, irreversibility and the ensuing evolution is not even understood in the time-incremental context. All further considerations are purely speculative at this juncture.

In particular, global minimality, which, as we saw earlier in Paragraph 4.2.3, entails relaxation even at the initial time, seems out of reach, because of our poor understanding of the interplay between relaxation and irreversibility. An attempt at reconciling relaxation and irreversibility was recently made in (Francfort and Garroni, 2006) and in (Dal Maso et al., 2006) in the much more pliant contexts of damage evolution and plasticity with softening respectively.

6. Path

Path, or rather the crack path, is in our view a byproduct of the time-continuous evolution. The weak variational evolution automatically delivers a crack path $\Gamma(t)$ for the time interval of study $[0, T]$. Since the only available existence result (Theorem 5.5) comes from the consideration of a global minimality criterion, together with a Griffith type surface energy, it only makes sense to discuss the path in that setting. This is the goal of this section.

An example of such a path was shown in the evolution computed at the end of Paragraph 5.1.4. Unfortunately, as mentioned before, we come woefully short on the issue of regularity of the obtained path. In the context of image segmentation (see Subsection 2.5), the existence result for the functional introduced in (Mumford and Shah, 1989) was obtained by (De Giorgi et al., 1989). It is based on the following regularity statement,

$$\mathcal{H}^1(\overline{S(\varphi_g)}\backslash S(\varphi_g)) = 0.$$

This result has been duplicated by Bourdin in his Ph.D. Thesis (Bourdin, 1998) for the weak discrete evolution described in Paragraph 5.1.1. He showed that $\mathcal{H}^1\left(\overline{\Gamma^n_{i+1}}\backslash\Gamma^n_{i+1}\right) = 0$ where Γ^n_{i+1} is the crack defined in the weak discrete evolution (Wde) of Paragraph 5.1.1 at time t^n_{i+1}. It is then a simple task to conclude that the pair $\left(\varphi^n(t), \overline{\Gamma^n(t)}\right)$ is a solution to the strong discrete evolution (Sde). The computed crack – an output of computations based on the weak discrete evolution – inherits "smoothness"; in other words, the components of the field $\varphi^n(t)$ are in $W^{1,p}\left(\Omega\backslash\overline{\Gamma^n(t)}\right)$.

For want of a similar regularity result at the time-continuous level, the closure of the theoretical crack, whose existence is shown through Theorem 5.5, could potentially be much bigger than the crack itself.

Besides, the evolution fails to assert uniqueness of the path, and, as in buckling, uniqueness should not be generically expected. Note that, in the context of image segmentation, uniqueness is but a conjecture even for the simplest geometries. Consider for example the functional

$$F(\varphi; A) := \int_A |\nabla\varphi|^2 \, dx + \mathcal{H}^1(S(\varphi) \cap A).$$

Then φ is said to be a *global minimizer* for F on $\mathbb{R}^2$ iff $F(\varphi; A) \leq F(\psi; A), \forall\psi$ with $\{\psi \neq \varphi\} \subset\subset A$, and $\forall A$ open. It is still a conjecture (due to De Giorgi) that $\varphi(r, \theta) := \sqrt{2r/\pi}\sin(\theta/2)$, with $\theta \in (-\pi, \pi)$, is the unique *global minimizer* for F (see (Bonnet and David, 2001) for the proof that φ is a global minimizer).

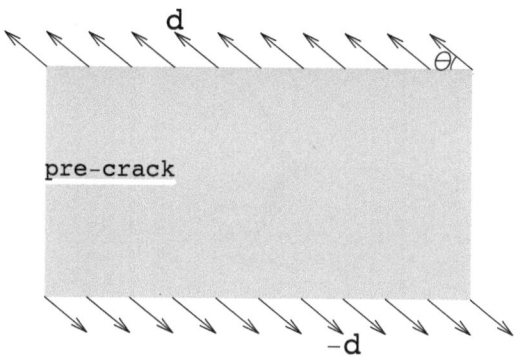

Figure 6.1. Pre-cracked sample

Summing up, the weak variational evolution delivers a crack path, in contrast to the Griffith formulation which postulates the crack path, or, in its post-modern version, imports additional ingredients of debatable universality for path prediction like the conflicted crack branching criteria (that of maximal energy release, still called $G_{\max}$, versus that of mode I propagation, still called $K_{II} = 0$). The predicted crack path may not be smooth or unique. It is however readily amenable to numerics through the weak discrete evolution. Lacking a better grasp on the theoretical properties of the path(s), "we should be apprehensive and cautious, as if on the brink of a deep gulf, as if treading on thin ice"[11] when attempting to weigh on the outstanding $G_{\max}$ versus $K_{II} = 0$ debate. And cautious we will be, contenting ourselves with a simplistic computation taken from (Bourdin et al., 2000) which demonstrates that branching of the path generated by the weak variational evolution may occur.

A pre-cracked 2d rectangular elastic plate is subject to a displacement load d as in Figure 6.1. The pre-crack is parallel to the horizontal sides of the rectangle. The angle θ that the displacement load makes with the horizontal line is set to a given value for each computation, while the intensity of the displacement is monotonically increased.

The next figures show the direction of the add-crack, whenever it appears, for a given value of the displacement angle θ.

In Figure 6.2, pure mode I propagation is observed, as expected.

In Figure 6.3, the crack branches at an angle which increases as θ decreases.

In Figure 6.4, two add-crack branches appear numerically. Here, when θ is small enough (and thus when the experiment gets closer and closer to a mode II experiment), the crack forks, which is not physical;

---

[11] Confucius – The Analects – VIII. 3. (191)

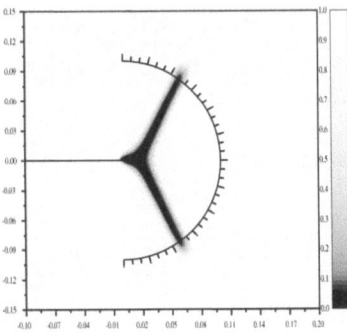

Figure 6.2. Mode I loading; $\theta = \pi/2$

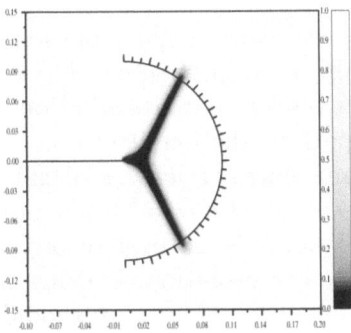

Figure 6.3. Mixed mode loading; $7\pi/180 < \theta < \pi/2$

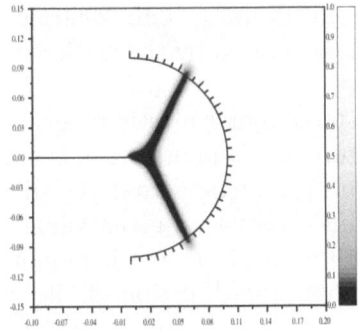

Figure 6.4. Mixed mode loading; $7\pi/180 > \theta$

indeed, interpenetration occurs along the upper branch of the fork. This is because our model does not forbid interpenetration as it should, as already emphasized in Subsection 2.1.

It is clear that such a numerical experiment, while providing evidence of branching, cannot be precise enough to allow for even a conjecture as to the nature of the relationship between θ and the branching angle.

So our contribution to the outstanding path issues is minimal at best. From an engineering standpoint, we do provide a computable path, which should then be compared to experiments. Branching does occur numerically, but cannot be quantified at present.

We complete this section with a multi-cracking example which further demonstrates the flexibility of the proposed method. Consider a cylindrical composite domain of length L (along the x-axis) and circular cross-section of area S made of an elastic unbreakable core (with cross-sectional area cS, $0 < c < 1$, Young's modulus E_f, and Poisson's ratio ν_f) of circular-cross section, surrounded by a brittle elastic annulus (with cross-sectional area $(1 - c)S$, Young's modulus E_m, Poisson's ratio ν_m, and fracture toughness k_m).

The annulus is perfectly bonded to the core. The sample is clamped at its far left cross-section and submitted to a monotonically increasing displacement load δ at its far right cross-section, as illustrated in Figure 6.5.

Figure A on page 100 shows snapshots of the deformed state of the cylinder at increasing values of the parameter δ. These are very large 3d-computations obtained using the method presented in Section 8. The Poisson coefficients of both material are assumed to equal to .2. The ratio E_f/E_m is 10. The diameter of the cylinder is 2, that of the inner core 1. The length of the domain is $L = 20$. The computations are performed on $1/4$ of the domain, so as to enforce symmetry of the solution with respect to the planes xy and xz. As the load increases, annular cracks appear brutally. The crack planes are equidistant, and the cracks seem to propagate from the end pieces toward the middle of the cylinder. However, the interval of loads within which this happens is so small that it is hard to identify the relation between the number of cracks and the load from the numerical experiments.

A theoretical derivation of the main features of the observed evolution from the only consideration of the weak variational evolution is a daunting task. As a first step in that direction, we provide below a partial analysis which demonstrates that, at the expense of a few educated

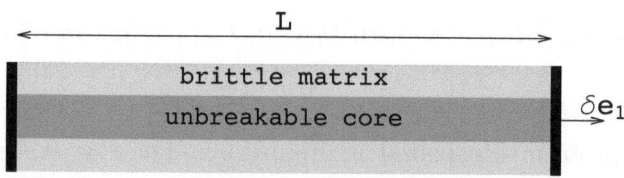

Figure 6.5. Composite tube in traction

guesses (see (6.1), (6.2)), several non trivial features of the numerically observed pattern – its periodicity, the increase in the number of cracks with the loads, ... – seem to be a natural consequence of the variational evolution. In truth, the theoretical/numerical fit is far from perfect as witnessed by e.g. the lack of periodicity of the crack distribution for small δ's, so that, barring stronger evidence, we could have embarked "on a fool's errand from the outset"[12].

In any case, we assume first that, throughout the purported evolution,

$$\textit{The only possible crack states are annular cracks} \atop \textit{with same thickness as that of the matrix.} \qquad (6.1)$$

REMARK 6.1. Our foolishness might be plain for all to see, were de-bonding to prove energetically more convenient than the annular cracking process envisioned in (6.1). A more detailed study, not undertaken in this tract, would strive to energetically confront those two obvious competitors and show that annular cracks are at first the favored mechanism, while de-bonding will take over for large enough values of the "load" δ. This is intuitively plausible because de-bonding of a region along the core between two annular cracks completely shields the said region from bulk energy, with an energetic price proportional to the length (along the x-axis of that region). Quantification of this remark is quite a challenge.

Assume n cracks with a first crack Γ_1 at a distance l_0 from the far left, the subsequent cracks Γ_{i+1} being at a distance l_i from its predecessor; l_n is the distance from Γ_n to the section $\{x = L\}$. The Griffith surface energy is $n(1-c)Sk_m$, while the spacings must satisfy

$$L = \sum_{0}^{n} l_i.$$

The computation of the elastic energy is the main obstacle to a rigorous analysis. Denoting by u the displacement field throughout the sample and by σ the Cauchy stress, we assume that, on the fiber cross-section alined with the cross-section of crack Γ_i,

$$u_1 := u \cdot e_1 = U_i(cst.), \text{ with } U_0 = 0, \ U_{n+1} = \delta; \quad \sigma e_1 \parallel e_1. \quad (6.2)$$

So, generically, we denote by $\frac{1}{2}\mathcal{A}(l)$ the elastic energy associated with the problem illustrated in Figure 6.6. For $i \neq 0, n$ the elastic

---

[12] Lord Byron – Correspondence with the Hon. Augusta Leigh

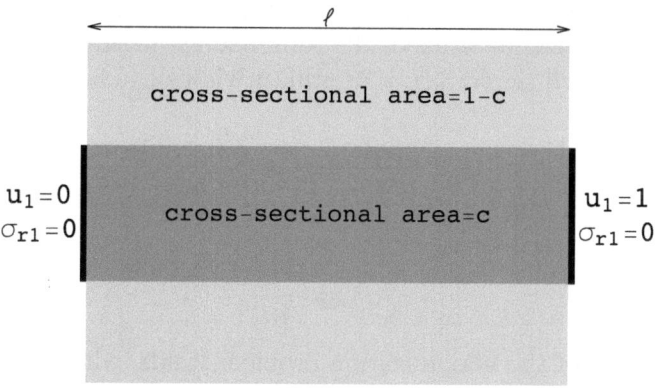

$$\ell$$

cross-sectional area=1-c

$u_1 = 0$
$\sigma_{r1} = 0$

cross-sectional area=c

$u_1 = 1$
$\sigma_{r1} = 0$

Figure 6.6. Cross-section between to cracks

energy associated with the piece of the sample located between Γ_i and Γ_{i+1} is, by reason of homogeneity, $\sqrt{S}/2\mathcal{A}(l_i/\sqrt{S})(U_{i+1} - U_i)^2$, while for $i = 0, n$, it is, by reason of symmetry, $\sqrt{S}\mathcal{A}(2l_0/\sqrt{S})U_1^2$, resp. $\sqrt{S}\mathcal{A}(2l_n/\sqrt{S})(\delta - U_n)^2$.

The reader should manipulate units with caution in the example discussed here because 1 has the dimension of a surface area in Figure 6.6. Also note that, except when $\nu = 0$, it is not so that $u_2 = 0$ at $x_2 = l/2$ in Figure 6.6, so that we may have underestimated the elastic energy associated with the first material segment $(i = 0)$.

For a fixed number n of cracks with set spacings $l_0, l_1, ..., l_n$, the elastic energy for the problem is obtained by minimizing

$$1/2 \sum_{i=1}^{n-1} \mathcal{A}(l_i/\sqrt{S})(U_{i+1} - U_i)^2 + \mathcal{A}(2l_0/\sqrt{S})U_1^2 + \mathcal{A}(Sl_n/\sqrt{S})(\delta - U_n)^2,$$

among all $U_1, ..., U_n$ (recall that $U_0 = 0$, $U_{n+1} = \delta$). We set $\varepsilon_i := U_{i+1} - U_i$, $i \neq 0, n$, and $\varepsilon_0 := U_1$, $\varepsilon_n := \delta - U_n$. Then

$$\sum_{i=0}^{n} \varepsilon_i = \delta,$$

and we must minimize

$$1/2 \sum_{i=1}^{n-1} \mathcal{A}(l_i/\sqrt{S})\varepsilon_i^2 + \mathcal{A}(2l_0/\sqrt{S})\varepsilon_0^2 + \mathcal{A}(2l_n/\sqrt{S})\varepsilon_n^2.$$

The minimum value of the elastic energy is given by

$$E(n; l_0, ..., l_n) = \frac{\delta^2 \sqrt{S}}{2} \left(\frac{1}{2\mathcal{A}(2l_0/\sqrt{S})} + \frac{1}{2\mathcal{A}(2l_n/\sqrt{S})} + \sum_{i=1}^{n-1} \frac{1}{\mathcal{A}(l_i/\sqrt{S})} \right)^{-1}. \quad (6.3)$$

Now, for a given number n of cracks, the surface energy is fixed. Thus, to minimize, at n fixed, the total energy, it suffices to minimize $E(n; \cdot)$ among all $(l_0, l_1, ..., l_n)$, or still, in view of (6.3), to compute

$$\max_{l_0, ..., l_n} \left\{ \frac{1}{2\mathcal{A}(2l_0/\sqrt{S})} + \frac{1}{2\mathcal{A}(2l_n/\sqrt{S})} + \sum_{i=1}^{n-1} \frac{1}{\mathcal{A}(l_i/\sqrt{S})} : \sum_{i=0}^{n} l_i = L \right\}.$$

Set

$$\mathcal{S}(l) := \frac{1}{\mathcal{A}(l)}.$$

The variation of the maximization problem yields, with classical notation,

$$\begin{cases} \sum_{i=0}^{n} dl_i = 0 \\ dl_0 \mathcal{S}'(2l_0/\sqrt{S}) + dl_n \mathcal{S}'(2l_n/\sqrt{S}) + \sum_{i=1}^{n-1} dl_i \mathcal{S}'(l_i/\sqrt{S}) = 0. \end{cases}$$

Thus,

$$\mathcal{S}'(2l_0/\sqrt{S}) = \mathcal{S}'(l_1/\sqrt{S}) = ... = \mathcal{S}'(l_{n-1}/\sqrt{S}) = \mathcal{S}'(2l_n/\sqrt{S}). \quad (6.4)$$

We lack at present a good grasp of the properties of $\mathcal{A}$ as a function of l. Elementary Reuss-Voigt type bounds (Landau and Lifschitz, 1991) immediately yield

$$\frac{l}{(cE_f + (1-c)E_m)} \leq \mathcal{S}(l) = \frac{1}{\mathcal{A}(l)} \leq l \left(\frac{c}{E_f} + \frac{(1-c)}{E_m} \right),$$

while an asymptotic analysis of the cell problem defining $\mathcal{A}(l)$ would demonstrate that, at least when $\nu_f = \nu_m =: \nu$,

$$\mathcal{A}(l) = \frac{E_f c}{l} + O(1) \text{ near } 0$$

$$\mathcal{A}(l) = \frac{E_f c + E_m(1-c)}{l} - \frac{K}{l^2} + o(l^{-2}) \text{ near } \infty,$$

where K depends on E_f, E_m, ν, c; see (Abdelmoula and Marigo, 2000), (Bilteryst and Marigo, 2003) for a detailed study of the asymptotic properties of $\mathcal{A}(l)$ near $l = \infty$ in a similar setting. Thus $\mathcal{S}(l)$ monotonically increases in l from 0 to ∞, with $1/(E_f c)$ as slope at 0 and

$$\mathcal{S}(l) \approx \frac{K}{(E_f c + E_m(1-c))^2} + \frac{l}{E_f c + E_m(1-c)}, \quad l \to \infty, \quad (6.5)$$

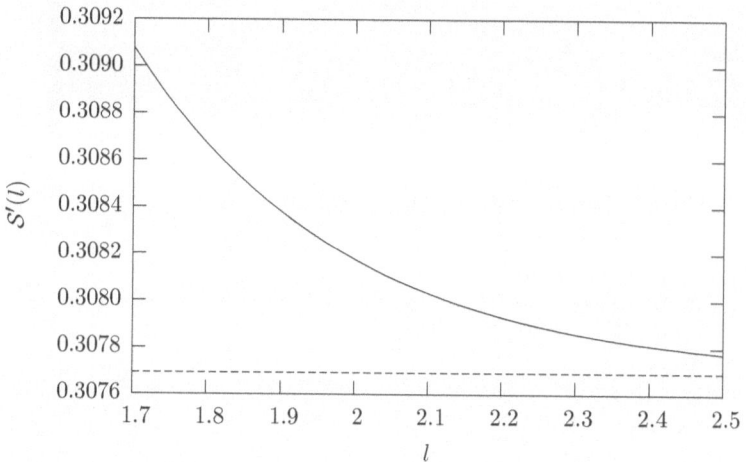

Figure 6.7. Concavity of $\mathcal{S}$ for $\dfrac{E_f}{E_m} = 10, \nu = 0.2, c = 0.25$

hence $1/(E_f c + E_m(1 - c))$ as asymptotic value for $\mathcal{S}'(l)$.

Numerical evidence for its part – see Figure 6.7 – suggests that, for large enough values of E_f/E_m, $\mathcal{S}(l)$ is a strictly concave function of l.

Then, by virtue of (6.4), the cracks *must be periodically distributed* except at the end sections, that is $2l_0 = l_1 = ... = 2l_n$! This result agrees with the experimental results of (Garrett and Bailey, 1977) on composites.

The minimum value of the elastic energy in (6.3) becomes

$$\mathcal{E}(n) = \frac{\sqrt{S}}{2n}\mathcal{A}\Big(\frac{L}{n\sqrt{S}}\Big)\delta^2 + n(1 - c)Sk_m,$$

which should be minimized in n, for fixed δ. Set $\eta := n\sqrt{S}/L$, so that

$$\mathcal{E}(n) = \mathcal{E}_0(\eta) := \frac{S}{2L}\mathcal{B}(\eta)\delta^2 + \eta(1 - c)\sqrt{S}Lk_m,$$

with

$$\mathcal{B}(\eta) := \frac{1}{\eta}\mathcal{A}\Big(\frac{1}{\eta}\Big).$$

Note that, near $\eta = 0$, that is for a small number of transverse cracks, $\mathcal{E}_0$ becomes, by virtue of (6.5),

$$\mathcal{E}_0(\eta) = \frac{S}{2L}(E_f c + E_m(1 - c))\delta^2 + \Big((1 - c)\sqrt{S}Lk_m - \frac{KS\delta^2}{2L}\Big)\eta + o(\eta).$$

The convexity properties of $\mathcal{B}$ become the determining feature of the evolution. For all tested values of the parameters, $\mathcal{B}$ is found to be decreasing and convex. Thus, initiation of the transverse cracking process

Figure 6.8. Transverse periodic cracking

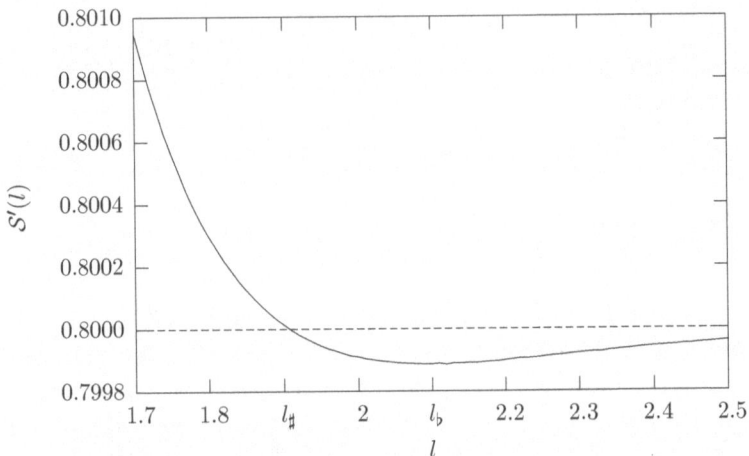

Figure 6.9. Non-concavity of $\mathcal{S}$ for $\dfrac{E_f}{E_m} = 2, \nu = 0.2, c = 0.25$

will occur when $\delta = \delta_i$ with

$$\frac{\delta_i}{L} := \sqrt{\frac{2k_m(1-c)}{K\sqrt{S}}}.$$

Then, the crack density will increase (cf. Figure 6.8) with increasing δ according to

$$\eta = (\mathcal{B}')^{-1}\left(-K\left(\frac{\delta_i}{\delta}\right)^2\right).$$

Numerical evidence – see Figure 6.9 – also suggests that, for values of E_f/E_m close enough to 1, $\mathcal{S}(l)$ is not a concave function of l.

Assume, as is the case on Figure 6.9, that $\mathcal{S}'$ decreases on $(0, l_\flat)$, crossing the asymptote at $l_\#$, then grows back to the asymptote. Then a more detailed study, under the numerically tested assumption that $\mathcal{B}$ is still decreasing and convex, would show the following evolution: an elastic phase, up to a value δ_0, a periodic cracking process with a fixed period p on an increasing volume fraction of the cylinder when

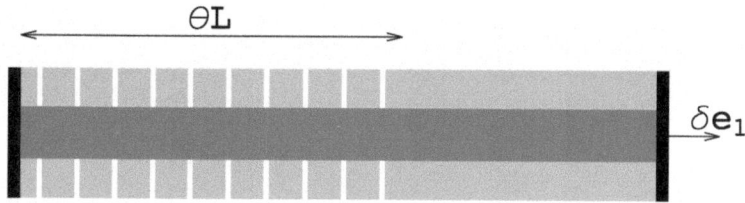

Figure 6.10. Transverse periodic cracking with period $l_{\#}\sqrt{S}$ on part of the plate

$\delta_0 < \delta < \delta_1$, and finally a periodic cracking process with an increasing crack density when $\delta > \delta_1$; the parameters δ_0, δ_1, p are explicit functions of the mechanical and geometric parameters, as well as of $l_{\#}, l_b$; see Figure 6.10.

The reader is spared the detailed derivation of the evolution in the non-concave case.

At the close of this section, we hope to have convincingly argued that the proposed variational approach "stands a ghost of a chance" when it comes to capturing complicated crack paths.

In the spirit of the previous computation, we ran the following variant. In Figure B on page 101, the total length of the cylinder is now 30 while the material parameters are those of the previous experiment. We wished to break the implicit symmetry hypothesis on the cracking process. To this effect, we created a half disk-shaped hairline crack of radius .4 centered on the outer edge of the brittle cylinder, along its middle cross-section. As the load increases, the existing crack grows smoothly through the cross-section of the brittle cylinder, until it reaches the interface of the inner reinforcement at which point it simultaneously grows along the interface, and along a helix-shaped path (see the top 2 figures in Figure B). As the load increases, the behavior changes. The following cycle repeats multiple times along the left side, then the right side of the domain: brutal propagation describing nearly one revolution along an helix-shaped path, then stagnation (or very slow growth). The jump in crack length between the third and fourth frames in Figure B, page 101, corresponds to the brutal phase in one cycle. In the final configuration, an helix–shaped crack spans the entire length of the domain.

Increasing the load beyond what is depicted here would result in the crack propagating along the matrix-reinforcement interface until total de-bonding, as in Figure 3 in (Bourdin, 2006).

Needless to say, we did not attempt to study the convexity of $\mathcal{S}(l)$ in this case!

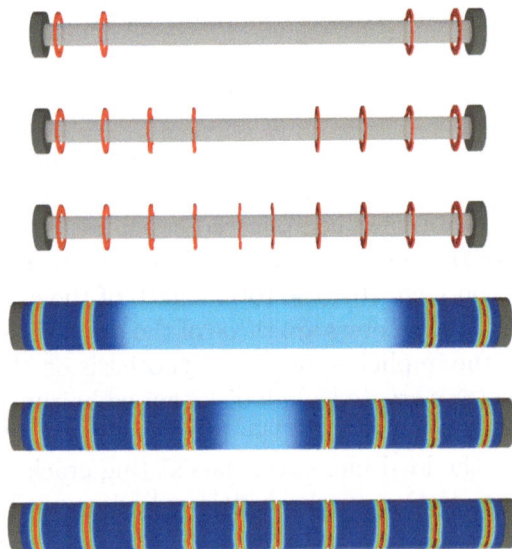

Figure A. Multi-cracking example of Section 6, $E_f/E_m = 10$, $\delta = 31.7$, 32.4, 35.4. The *top three figures* represent the geometry of the crack set. The *bottom three figures* represent the domain in its deformed configuration. The *color coding* represents the smeared crack field v that replaces the actual crack in the numerical approximation (see Section 8 for details)

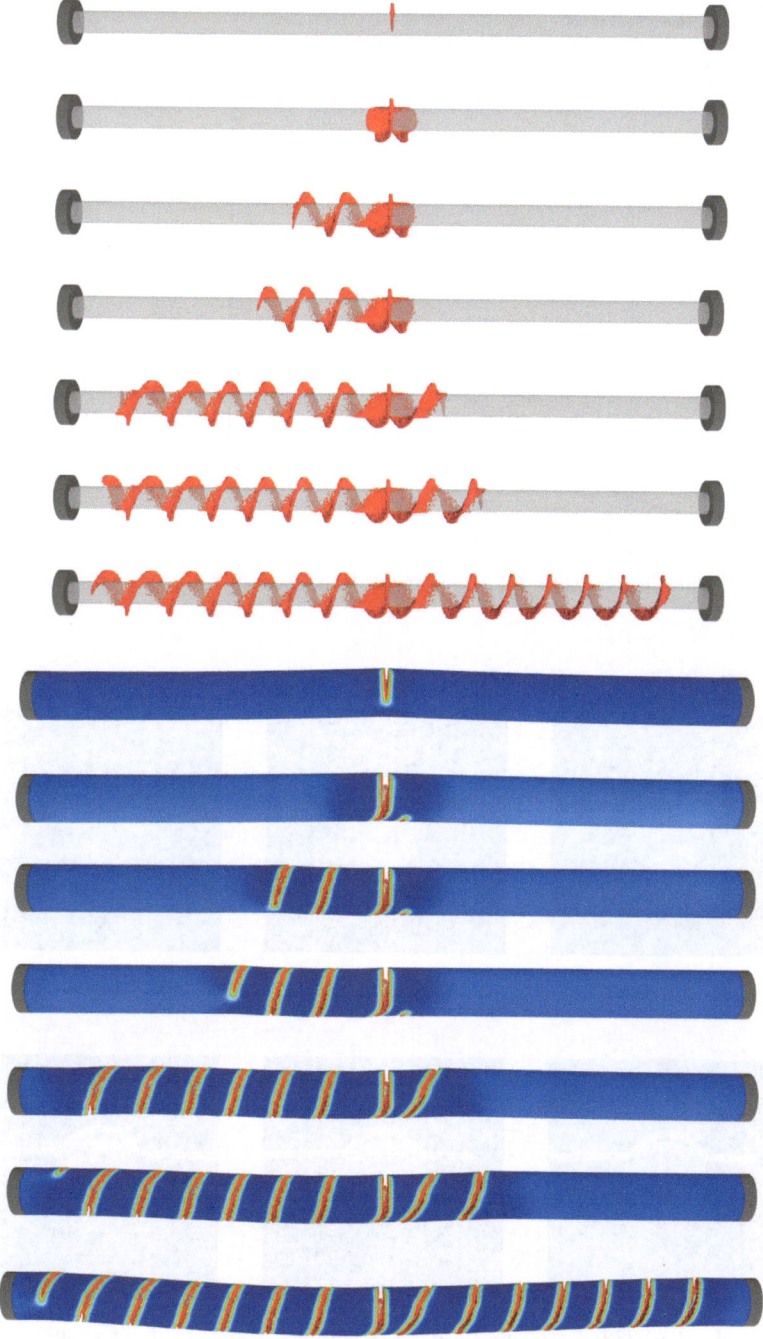

Figure B. Non symmetric multi-cracking example of Section 6, $E_f/E_m = 1$, $\delta = 28$, 42, 45, 48, 65.5, 69.5, 86.5

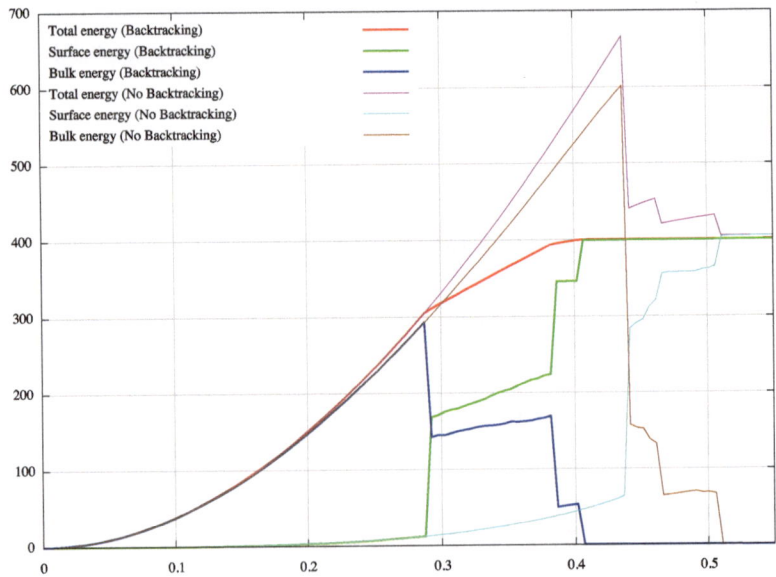

Figure C. Energies vs. time for the plate of Paragraph 8.3.3

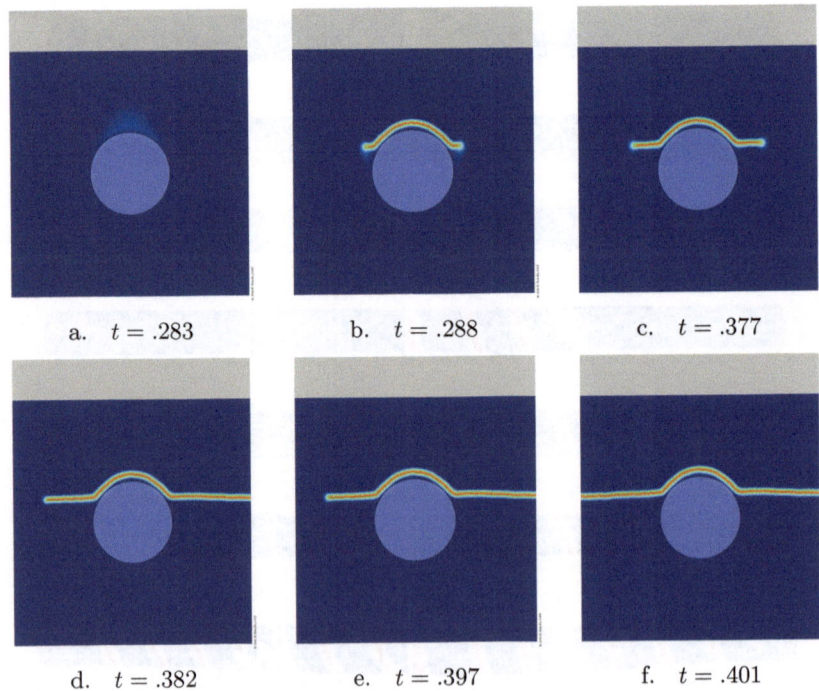

a. $t = .283$ b. $t = .288$ c. $t = .377$

d. $t = .382$ e. $t = .397$ f. $t = .401$

Figure D. Snapshots of the crack evolution for the plate of Paragraph 8.3.3

7. Griffith vs. Barenblatt

Mechanical folklore has it that, for a given brittle sample, cohesive models à la Barenblatt will behave asymptotically like Griffith's model as the internal length shrinks to 0. By internal length, we mean the ratio between the fracture toughness and the yield stress; see e.g. Figure 2.3 in Subsection 2.6. Provided that cohesive forces are only triggered near the crack tip, similar views were already espoused by (Griffith, 1920), page 166 : "it may therefore be said that the application of the mathematical theory of elasticity on the basis that the crack is assumed to be a traction-free surface, must give the stresses correctly at all points of the body, with the exception of those near the ends of the crack. In a sufficiently large crack the error in the strain energy so calculated must be negligible."

Our purpose in what follows is to quantify this within our framework of choice, the variational framework. We visit this issue in the context of global minimality and report on Giacomini's significant contribution (Giacomini, 2005b). We forego a general investigation of local minimality because of the current lack of any kind of meaningful results, but refer the reader to (Marigo and Truskinovsky, 2004) in the case of a pull-out problem, or to Section 9 in the context of fatigue.

At first glance, the investigation of the asymptotic behavior of the cohesive variational evolution may seem oxymoronic in view of our failure – reported in Subsections 4.2, 5.2 – to secure a meaningful notion of evolution in the cohesive setting. This is indeed so, if we insist on viewing the existence of a time-continuous evolution in the cohesive case as a prerequisite.

The viewpoint espoused in (Giacomini, 2005b) is slightly different. We propose to describe his work in this subsection. Giacomini starts, as we did in Paragraph 5.1.1 with a time discretization of a hypothetical relaxed variational evolution for the cohesive model in the global minimality context that we introduced in Subsection 2.6. In other words, taking the simplest available framework, that of anti-plane shear in linearized elasticity, he considers a finite set of energies constructed after the model relaxed energy (4.19) obtained in Paragraph 4.2.3, each element of this set corresponding to a time in the discretization of the interval $[0, T]$ of study.

As in Subsection 5.1, we assume throughout this subsection that the only load is a displacement $g(t)$ defined on $\partial_d \Omega$, or rather, as we saw earlier in Subsection 2.5, on $\mathbb{R}^2 \backslash \overline{\Omega}$. Suppose that φ_j^n and Γ_j^n are known for $j = 0, \ldots, i - 1$, and define, in the notation of Subsection

2.6, Paragraph 4.2.3 (see (4.20)), φ_i^δ to be a minimizer for

$$\min \left\{ \int_\Omega \hat{W}(\nabla\varphi)\,dx + \int_{(S(\varphi)\backslash\partial_s\Omega)\cup\Gamma_{i-1}^n} \kappa\left(\left|[\varphi]\right| \vee \psi_{i-1}^n\right)\,d\mathcal{H}^1 + \sigma_y|C(\varphi)| :\right.$$

$$\left.\varphi = g(t_{i+1}^n) \text{ on } \partial_d\Omega\backslash(S(\varphi)\cup\Gamma_{i-1}^n)\right\} \tag{7.1}$$

where $\psi_{i-1}^n := |[\varphi_0^n]| \vee \cdots \vee |[\varphi_{i-1}^n]|$ and set $\Gamma_i^n := \Gamma_{i-1}^n \cup (S(\varphi_i^n)\backslash\partial_s\Omega)$. Note that, in this approach, the irreversibility constraint is encoded in ψ_{i-1}^n, and it consists – as first introduced through (2.27) in Subsection 2.6 and then revisited in Subsection 5.2 – in assuming that the surface energy increases only when the crack lip displacement increases. Other choices, such as that of a cumulative increment, could be made (see Subsection 5.2); that latter choice will be used in going from fracture to fatigue in Section 9.

Although the functional introduced in (7.1) is close to that in (4.19), it differs on two grounds: first, irreversibility has been introduced via ψ_{i-1}^n; then boundary conditions have been imposed on $\partial_d\Omega$. Consequently, the relaxation result of (Bouchitté et al., 1995) does not directly apply and a first task consists in showing, see (Giacomini, 2005b), Section 9, that φ_i^n exists and that it is such that

$$\int_\Omega \hat{W}(\nabla\varphi_i^n)\,dx + \int_{\Gamma_i^n} \kappa(\psi_i^n)\,d\mathcal{H}^1 + \sigma_y|C(\varphi_i^n)|$$

$$= \inf_\varphi \left\{ \int_\Omega W(\nabla\varphi)\,dx + \int_{(S(\varphi)\backslash\partial_s\Omega)\cup\Gamma_{i-1}^n} \kappa\left(\left|[\varphi]\right| \vee \psi_{i-1}^n\right)\,d\mathcal{H}^1 \right\},$$

or, as explained in Paragraph 4.2.3, that, at each time step,

$$\mathcal{E}_i^*(\varphi) := \int_\Omega \hat{W}(\nabla\varphi)\,dx + \int_{(S(\varphi)\backslash\partial_s\Omega)\cup\Gamma_{i-1}^n} \kappa\left(\left|[\varphi]\right| \vee \psi_{i-1}^n\right)\,d\mathcal{H}^1$$

$$+ \sigma_y|C(\varphi)|$$

is the relaxed energy (for the weak-* topology in $BV(\Omega)$) of

$$\mathcal{E}_i(\varphi) := \int_\Omega W(\nabla\varphi)\,dx + \int_{(S(\varphi)\backslash\partial_s\Omega)\cup\Gamma_{i-1}^n} \kappa\left(\left|[\varphi]\right| \vee \psi_{i-1}^n\right)\,d\mathcal{H}^1.$$

The BV-bound on quasi-minimizers for $\mathcal{E}_i$ will be easily obtained, provided that $W(F) \geq \mathcal{C}(|F| - 1/\mathcal{C})$.

So, at this point, we have derived a discrete cohesive relaxed weak variational evolution in the global minimality setting!

Now is the time to introduce the varying parameter for the asymptotic analysis, namely the internal length. With $h \nearrow \infty$, we replace $\hat{W}$ and κ by, respectively $\hat{W}_h$ and κ_h given by

$$\hat{W}_h(F) := \inf\{W(G) + \sigma_y h|H|; \ G + H = F\},$$

and
$$\kappa_h(s) := \kappa(hs),$$

while we replace the Cantor contribution $\sigma_y|C(\varphi)|$ – that related to "diffuse cracking", see Subsection 2.6 – by $\sigma_y h|C(\varphi)|$. The corresponding fields minimizing (7.1) are denoted by φ_{hi}^n, ψ_{hi}^n, Γ_{hi}^n.

We then specialize the discretization parameter n to be of the form $n(h) \nearrow^h \infty$. Assuming decent regularity (see (5.1)) on the boundary displacement $g(t)$ as in Subsection 5.1, we define the piecewise in time fields

$$\begin{cases} \varphi_h(t) := \varphi_{hi}^{n(h)} \\ \Gamma_h(t) := \Gamma_{hi}^{n(h)} \quad \text{on } [t_i^{n(h)}, t_{i+1}^{n(h)}), \text{ and, for } i = -1, \Gamma_{h(-1)}^n := \Gamma_0. \\ \psi_h(t) = \psi_{hi}^{n(h)} \end{cases}$$

We obtain the following a priori bounds:

$$\|\varphi_h(t)\|_{BV(\Omega)} \le \mathcal{C}, \quad |C(\varphi_h(t))| \le \mathcal{C}/h, \qquad (7.2)$$

and, providing that the energy density $W(F)$ behaves like $|F|^p$,

$$|\nabla\varphi_h(t)| \text{ is equi-integrable, uniformly on } [0,T]. \qquad (7.3)$$

An energy upper bound similar to (5.4) can also be derived. It reads – with notation borrowed from (5.4) – as

$$E_h(t) := \int_\Omega \hat{W}_h(\nabla\varphi_h(t)) \, dx + \int_{\Gamma_h(t)} \kappa_h\left(\left|[\psi_h(t)]\right|\right) d\mathcal{H}^1 + \sigma_y h|C(\varphi_h(t))|$$

$$\le E_h(0) + \int_0^{\tau_h(t)} \int_\Omega \frac{\partial W}{\partial F}(\nabla\varphi_h(s)) \cdot \nabla\dot{g}(s) \, dx \, ds + O(1/h).$$

Then, a variant of Ambrosio's compactness theorem (see (2.24)), proved in (Giacomini, 2005b) establishes in particular that if $\varphi(t)$ is the weak limit of (a time-dependent subsequence of) $\varphi_h(t)$, then $\varphi(t) \in SBV(\Omega)$.

Following a path similar to that in Subsection 5.1, the next step consists in showing that $\varphi(t)$ satisfies the global minimality statement (Ugm) in the weak variational evolution. In Paragraph 5.1.2, this was achieved with the help of two essential ingredients: the jump transfer result, Theorem 5.1, and a meaningful definition of a limit crack through σ^p-convergence, Definition 5.2 and Theorem 5.4. This same path is followed in (Giacomini, 2005b).

The jump transfer theorem is adapted to the situation at hand by replacing

$$\limsup_h \mathcal{H}^1 \lfloor A\left(S(\zeta_h)\backslash S(\varphi_h)\right) \le \mathcal{H}^1 \lfloor A\left(S(\zeta)\backslash S(\varphi)\right)$$

in that theorem with

$$\limsup_h \left[\int_{A\cap(S(\zeta_h)\cup S(\varphi_h))} \kappa_h(|[\zeta_h]| \vee |[\varphi_h]|)\, d\mathcal{H}^1 - \int_{A\cap(S(\varphi_h))} \kappa_h(|[\varphi_h]|)\, d\mathcal{H}^1 \right]$$

$$\leq \mathcal{H}^1 \lfloor A\, (S(\zeta)\backslash S(\varphi))\,,$$

while Definition 5.2 and Theorem 5.4 are correspondingly adapted (see Subsection 5.2 in (Giacomini, 2005b)).

From that point on, the argument follows closely that outlined in Subsection 5.1. The resulting theorem, stated in (Giacomini, 2005b) in the case where $W(F) = 1/2|F|^2$, but generalizable to the setting of Theorem 5.5 – at least in the anti-plane shear setting – is as follows:

THEOREM 7.1. *There exists a t-independent subsequence of $\{h \nearrow \infty\}$ – still denoted $\{h\}$ –and a weak quasi-static evolution pair $(\varphi(t), \Gamma(t))$ satisfying all conclusions of Theorem 5.5 such that, for all $t \in [0, T]$,*

- *$\varphi_h(t) \rightharpoonup \varphi(t)$ weak-* in $BV(\Omega)$;*

- *$\nabla\varphi(t) \rightharpoonup \nabla\varphi(t)$ weakly in $L^1(\Omega; \mathbb{R}^2)$;*

- *Every accumulation point χ of $\varphi_h(t)$ (in the weak-* topology of $BV(\Omega)$) is in $SBV(\Omega)$ and such that $S(\chi) \subset \Gamma(t)$, $\nabla\chi = \nabla\varphi(t)$;*

- *$E_h(t) \to E(t)$, the total energy associated with the evolution pair;*

- *$\int_\Omega W_h(\nabla\varphi(t))\, dx \to \int_\Omega W(\varphi(t))\, dx$; and*

- *$\int_{\Gamma_h(t)} \kappa_h(\psi_h(t))\, d\mathcal{H}^1 \to \mathcal{H}^1(\Gamma(t))$.*

The above result firmly anchors folklore in reality. As the size of the process zone shrinks, the time-discrete cohesive evolution – the admittedly pale substitute for a bona fide evolution in the cohesive setting – will converge (for a time-step which goes to 0 with the size of the process zone) to a weak evolution for the associated asymptotic Griffith state of the surface energy. A litigious reader might, rightfully, object to the arbitrary nature of the description of irreversibility/dissipation in this subsection. We will see in Section 9 below that the same is expected with a different choice for the dissipation, and it is sheer laziness that has prevented us from revisiting Giacomini's arguments in the latter setting.

8. Numerics and Griffith

At first glance, numerical implementation of the variational approach advocated in this tract is hopeless and "all goes wrong when our unhappy cause becomes connected with it. Strength becomes weakness, wisdom folly"[13] as the variational approach attempts to free the crack path because the "classical" numerical methods dealing with discontinuous displacement fields rely on some non-negligible amount of *a priori* knowledge of that path. This includes the extended finite element method and other enrichment-based variants. A proper discretization scheme for the total energy needs to both approximate potentially discontinuous displacement fields – and thus the position of their discontinuity sets – and lead to an accurate and isotropic approximation of the surface energy. Such a scheme does not easily accommodate cohesive finite element methods or discontinuous Galerkin methods. Note that this is partially addressed by a careful estimate of the anisotropy induced by the mesh in (Negri, 1999), (Negri, 2003) or still through the use of adaptive finite element methods (Bourdin and Chambolle, 2000).

Further, if the variational framework contends that it addresses crack initiation and crack propagation in a unified framework, the same should be true of the numerical method. In particular, methods based on considering energy restitution caused by small increments of existing cracks are ruled out. In view of Proposition 4.3, "small" cracks will never lead to descent directions for the global minimization of the total energy in the absence of strong singularities in the elastic field.

Non-convexity of the total energy is yet another major obstacle to overcome. The typical size of the discrete problems prohibits appeal to global or non-deterministic optimization techniques. As seen in details in previous sections, global minimization of the energy is an arguable postulate, but it is at present the only one theoretically suitable for a thorough investigation of any numerical implementation.

As mentioned in the Introduction, and as also suggested by the title of this section, the scope of the numerics does not extend beyond the Griffith setting. Indeed, as seen several times before, global minimization, the only numerically viable option, does not lead as of yet to a well understood evolution in the cohesive setting, so that any numerical incursion into the cohesive territory would be hazardous. Also, recalling the argument put forth at the start of Paragraph 4.1.1, it will then come to no surprise that the only loads considered throughout this section are displacement loads.

---

[13] Sir Walter Scott – Anne of Geierstein

The numerical method that will be described below finds, once again, its inspiration in the Mumford-Shah functional for image segmentation (see Subsection 2.5). The main ingredients were first introduced in the latter context in (Ambrosio and Tortorelli, 1990), (Ambrosio and Tortorelli, 1992), (Bellettini and Coscia, 1994), (Bourdin, 1998), (Bourdin, 1999), (Negri and Paolini, 2001) and later adapted to fracture in (Bourdin et al., 2000), (Giacomini and Ponsiglione, 2003), (Chambolle, 2004), (Chambolle, 2005), (Giacomini, 2005a), (Giacomini and Ponsiglione, 2006).

The method allows for an isotropic and mesh independent approximation of the total energy. It copes rather successfully with both initiation and propagation as seen through the various numerical experiments presented in Subsection 8.3. Like the actual variational model, it applies to the one, two, or three dimensional cases without alteration.

Finally, as first suggested in the Introduction, time dependence will be approached through time discretization, and all computations will be performed for a sequence of times $t_0 = 0 < t_1^n < \dots < t_{k(n)}^n = T$ with $k(n) \nearrow^{n} \infty$, $\Delta_n := t_{i+1}^n - t_i^n \searrow^{n} 0$. We will mostly drop the n-dependence, unless explicitly referring to the putative convergence of the time-discrete evolution to the time-continuous evolution.

8.1. NUMERICAL APPROXIMATION OF THE ENERGY

The essence of the numerical implementation is to be found in the concept of variational convergence. Specifically, the first step consists in devising a good approximation of the total energy in the sense of Γ–convergence. We refer the reader to (Dal Maso, 1993), (Braides, 2002) for a complete exposition of the underlying theory.

Consider a $\overline{\mathbb{R}}$-valued functional $\mathcal{F}$ defined over, say a metrizable topological space X, and a sequence $\mathcal{F}_\varepsilon$ of the same type. Then, $\mathcal{F}_\varepsilon$ Γ–converges to $\mathcal{F}$ as $\varepsilon \searrow 0$ iff the following two conditions are satisfied for any $u \in X$:

1. for any sequence $(u_\varepsilon)_\varepsilon \in X$ converging to u, one has

$$\liminf_{\varepsilon \to 0} \mathcal{F}_\varepsilon(u_\varepsilon) \geq \mathcal{F}(u);$$

2. there exists a sequence $(u_\varepsilon)_\varepsilon \in X$ converging to u, such that

$$\limsup_{\varepsilon \to 0} \mathcal{F}_\varepsilon(u_\varepsilon) \leq \mathcal{F}(u).$$

The interest of Γ–convergence from the standpoint of numerics lies in the following elementary theorem in Γ-convergence:

THEOREM 8.1. *If $\mathcal{F}_\varepsilon$ Γ–converges to $\mathcal{F}$ and u_ε^* is a minimizer of $\mathcal{F}_\varepsilon$ and if, further, the sequence u_ε^* is compact in X, then there exists $u^* \in X$ such that $u_\varepsilon^* \to u$, u^* is a global minimizer for $\mathcal{F}$, and $\mathcal{F}_\varepsilon(u_\varepsilon^*) \to \mathcal{F}(u^*)$.*

Stability of global minimizers under Γ–convergence is indeed a powerful numerical tool. Rather than attempting to minimize the total energy – thus having to reconcile discretization and discontinuous functions – we propose to construct, at each time step t_i, a family of regularized energies $\mathcal{E}_\varepsilon^i$ that Γ-converge to $\mathcal{E}^i$, the energy for the weak variational evolution at that time step (see (2.25)). In the footstep of (Ambrosio and Tortorelli, 1990), (Ambrosio and Tortorelli, 1992), we will approximate the potentially discontinuous field φ^i and its crack set Γ^i by two smooth functions. The implementation of the first time step, which is very close to that of the original approximation in the context of the Mumford-Shah functional, is presented in Paragraph 8.1.1. while Paragraph 8.1.2 shows how to account for irreversibility and approximate the weak discrete time evolution (Wde).

8.1.1. *The first time step*

Consider the first time step of the weak discrete evolution under the unilateral global minimality condition (Ugm). The irreversibility condition is trivially satisfied, so that it suffices to minimize the total energy

$$\mathcal{E}(\varphi) = \int_\Omega W(\nabla\varphi)dx + \mathcal{H}^1(S(\varphi))$$

with respect to any kinematically admissible φ. In all that follows, $\widetilde{\Omega}$ denotes a "large enough" open bounded set such that $\Omega \subset \widetilde{\Omega}$, and the Dirichlet boundary conditions are enforced on $\widetilde{\Omega}\backslash\bar{\Omega}$, not on $\mathbb{R}^2\backslash\bar{\Omega}$ because, as will be seen below, the computations are performed on that larger domain, and not only on Ω.

Following (Ambrosio and Tortorelli, 1990), (Ambrosio and Tortorelli, 1992), we introduce a secondary variable $v \in W^{1,2}(\widetilde{\Omega}\backslash\partial_s\Omega)$ and two small positive parameters ε, and $\eta_\varepsilon = o(\varepsilon)$, and define, for any kinematically admissible φ,

$$\mathcal{F}(\varphi, v) = \begin{cases} \int_\Omega W(\nabla\varphi)\,dx + k\mathcal{H}^{N-1}(S(\varphi)\backslash\partial_S\Omega) & \text{if } v = 1 \text{ a.e.} \\ +\infty & \text{otherwise,} \end{cases} \tag{8.1}$$

and

$$\mathcal{F}_\varepsilon(\varphi, v) = \int_\Omega (v^2 + \eta_\varepsilon)W(\nabla\varphi)\,dx + k\int_{\widetilde{\Omega}\backslash\partial_s\Omega}\left\{\frac{(1-v)^2}{4\varepsilon} + \varepsilon|\nabla v|^2\right\}\,dx. \tag{8.2}$$

In the anti-plane case, proving the Γ–convergence of $\mathcal{F}_\varepsilon$ to $\mathcal{F}$ is a simple adaptation of Ambrosio and Tortorelli's result (see (Bourdin, 1998)) while it is more involved in that of linearized elasticity (Chambolle, 2004). We limit the analysis to the former case. The proof of the lower inequality of Theorem 8.1 is technical and does not shed much light on the proposed numerical method. By contrast, the construction of an attainment sequence in Theorem 8.1 provides valuable insight and we propose to detail it, at least when the target is a mildly regular kinematically admissible field for $\mathcal{E}$. Actually, deriving the lim-sup inequality for minimizers can easily be seen to be no restriction. But for those, the mild regularity assumption below holds true, at least in anti-plane shear and energy densities of the form $|F|^p$ with $p > 1$.

It is thus assumed that $|\varphi(x)| \leq M$ for a.e. $x \in \widetilde{\Omega}$ and for some $M > 0$. By the maximum principle, this is equivalent to imposing a similar bound on the initial load because a simple truncation at level M of $|\varphi|$ will then decrease the energy. It is also assumed that φ is a solution to the minimization of $\mathcal{E}$ that satisfies

$$\mathcal{H}^1(\overline{S(\varphi)}) = \mathcal{H}^1(S(\varphi)). \tag{8.3}$$

For minimizers of the Mumford-Shah functional, this mild, albeit difficult regularity property was established in (De Giorgi et al., 1989). In the scalar-valued setting, the case of a certain class of convex bulk energies which includes $p > 1$-homogeneous energies was investigated in (Fonseca and Fusco, 1997). The regularity result was generalized to our setting in (Bourdin, 1998), at least for minimizers in anti-plane shear with a quadratic elastic energy density. The closure property (8.3) is not so clearly true in more general settings, and different approximation processes must be used in such cases; the interested reader is invited to consult e.g. (Braides, 2002).

The energy will be assumed quadratic in the field, i.e., $W(F) := 1/2\mu|F|^2$, although more general convex energies would be permitted. The given construction does not account for the Dirichlet boundary condition and the interested reader is referred to (Bourdin, 1998) for the corresponding technicalities. As a corollary, we may as well take $\widetilde{\Omega} \equiv \Omega$ in the construction of the attainment sequence that follows. In truth, we are just considering an approximation of the weak form (2.23) of the Mumford-Shah functional in the derivation. In the case of interest to us, i.e., that with Dirichlet boundary conditions on a part $\partial_d\Omega = \partial\Omega\backslash\partial_s\Omega$ it will be enough to reintroduce $\widetilde{\Omega}\backslash\partial\Omega_s$ in lieu of Ω in the second integral in (8.2).

Consider a kinematically admissible field φ – an element of $SBV(\Omega)$ – satisfying (8.3). Define

$$d(x) := \text{dist}(x, \overline{S(\varphi)}).$$

The volume of the area bounded by the s-level set of d is

$$\ell(s) := \left| \left\{ x \in \mathbb{R}^2 \, ; \, d(x) \le s \right\} \right|.$$

The distance function is 1-Lipschitz, *i.e.*, $|\nabla d(x)| = 1$ a.e., while, by the co-area formula for Lipschitz functions (see e.g. (Ambrosio et al., 2000)),

$$\ell(s) = \int_0^s \mathcal{H}^1 \left(\{ x \, ; d(x) = t \} \right) dt,$$

so that, in particular,

$$\ell'(s) = \mathcal{H}^1 \left(\{ x \, ; d(x) = s \} \right). \tag{8.4}$$

Also, see (Federer, 1969)-3.2.39,

$$\lim_{s \to 0} \frac{\ell(s)}{2s} = \mathcal{H}^1(S(\varphi)).$$

We choose α_ε such that $\alpha_\varepsilon = o(\varepsilon)$, $\eta_\varepsilon = o(\alpha_\varepsilon)$, which is possible since $\eta_\varepsilon = o(\varepsilon)$, and define the functions

$$v_\varepsilon(x) := \begin{cases} 0 & \text{if } d(x) \le \alpha_\varepsilon \\ 1 - \exp\left(-\dfrac{d(x) - \alpha_\varepsilon}{2\varepsilon} \right) & \text{otherwise,} \end{cases} \tag{8.5}$$

and

$$\varphi_\varepsilon(x) := \begin{cases} \dfrac{d(x)}{\alpha_\varepsilon} \varphi(x) & \text{if } 0 \le d(x) \le \alpha_\varepsilon \\ \varphi(x) & \text{otherwise.} \end{cases}$$

Note that it is easily seen that $\varphi_\varepsilon \in W^{1,2}(\Omega)$. Further, $\varphi_\varepsilon \to \varphi$ in $L^2(\Omega)$, and $v_\varepsilon \to 1$ almost everywhere. Since $v_\varepsilon \le 1$,

$$\int_\Omega \left(v_\varepsilon^2 + \eta_\varepsilon \right) |\nabla \varphi_\varepsilon|^2 dx \le \int_{d(x) \le \alpha_\varepsilon} \eta_\varepsilon |\nabla \varphi_\varepsilon|^2 dx + \int_{d(x) \ge \alpha_\varepsilon} (1 + \eta_\varepsilon) |\nabla \varphi|^2 dx.$$

Observe now that, for $d(x) \le \alpha_\varepsilon$, $\nabla \varphi_\varepsilon = (d(x)/\alpha_\varepsilon) \nabla \varphi + (1/\alpha_\varepsilon) \varphi \nabla d$, so, in view of the 1-Lipschitz character of d and of the L^∞-bound on φ,

$$\int_\Omega (v_\varepsilon^2 + \eta_\varepsilon) |\nabla \varphi_\varepsilon|^2 \, dx \le 2 \left(\eta_\varepsilon \int_{d(x) \le \alpha_\varepsilon} |\nabla \varphi|^2 \, dx + M^2 \frac{\eta_\varepsilon}{\alpha_\varepsilon^2} \ell(\alpha_\varepsilon) \right)$$
$$+ \int_{d(x) \ge \alpha_\varepsilon} (1 + \eta_\varepsilon) |\nabla \varphi|^2 dx.$$

Since $\int_\Omega |\nabla \varphi|^2 \, dx < \infty$, the first term in the parenthesis on the right hand side above converges to 0 as $\varepsilon \to 0$. Recalling that $\ell(\alpha_\varepsilon)/\alpha_\varepsilon = O(1)$, while $\eta_\varepsilon/\alpha_\varepsilon = o(1)$ permits one to conclude that the limit of the second term in that parenthesis also converges to 0 with ε. We conclude that

$$\limsup_{\varepsilon \to 0} \int_\Omega \left(v_\varepsilon^2 + \eta_\varepsilon \right) |\nabla \varphi_\varepsilon|^2 \, dx \leq \int_\Omega |\nabla \varphi|^2 \, dx. \tag{8.6}$$

Let us examine the surface energy term. Using once again the 1-Lipschitz character of d, together with the co-area formula, we get

$$\int_\Omega \left\{ \varepsilon |\nabla v_\varepsilon|^2 + \frac{(1 - v_\varepsilon)^2}{4\varepsilon} \right\} dx \leq \frac{\ell(\alpha_\varepsilon)}{4\varepsilon} + \int_{d(x) \geq \alpha_\varepsilon} \frac{1}{2\varepsilon} \exp(-\frac{d(x) - \alpha_\varepsilon}{\varepsilon}) dx$$

$$\leq \frac{\ell(\alpha_\varepsilon)}{4\varepsilon} + \frac{1}{2\varepsilon} \int_{\alpha_\varepsilon}^\infty \exp(-\frac{s - \alpha_\varepsilon}{\varepsilon}) \mathcal{H}^1 \left(\{ d(x) = s \} \right) ds. \tag{8.7}$$

Recalling (8.4),

$$\frac{1}{2\varepsilon} \int_{\alpha_\varepsilon}^\infty \exp(-\frac{s - \alpha_\varepsilon}{\varepsilon}) \mathcal{H}^1 \left(\{ d(x) = s \} \right) ds = \frac{e^{\frac{\alpha_\varepsilon}{\varepsilon}}}{2\varepsilon} \int_{\alpha_\varepsilon}^\infty e^{-s/\varepsilon} \ell'(s) \, ds$$

$$= \frac{e^{\frac{\alpha_\varepsilon}{\varepsilon}}}{2} \int_{\alpha_\varepsilon/\varepsilon}^\infty e^{-t} \ell'(t\varepsilon) \, dt. \tag{8.8}$$

Since $\ell'(0) = \lim_{s \to 0} \ell(s)/s = 2\mathcal{H}^1(S(\varphi))$, $\alpha_\varepsilon = o(\varepsilon)$ and $\int_0^\infty e^{-t} \, dt = 1$, insertion of (8.8) into (8.7) and application of Lebesgue's dominated convergence theorem yields

$$\limsup_{\varepsilon \to 0} \int_\Omega \left\{ \varepsilon |\nabla v_\varepsilon|^2 + \frac{(1 - v_\varepsilon)^2}{4\varepsilon} \right\} dx \leq \mathcal{H}^1(S(\varphi)). \tag{8.9}$$

Collecting (8.6), (8.9) gives the upper Γ–limit inequality.

REMARK 8.2. The form of the field v_ε in (8.5) may seem somewhat ad-hoc. It is not. The choice of the profile for the field v_ε is derived from the solution of an "optimal profile" problem (see (Alberti, 2000)). Consider, in e.g. 2d, a point x on the crack and a line orthogonal to the crack and passing through x, parameterized by the variable s. Consider the restriction of the regularized surface energy to this line

$$\mathcal{F}_{\varepsilon,x}(s) = k \int_0^\infty \left\{ \frac{(1 - v(s))^2}{4\varepsilon} + \varepsilon |v'(s)|^2 \right\} ds.$$

Then the profile

$$v_\varepsilon(s) = 1 - \exp\left(-\frac{(s - \alpha_\varepsilon)}{2\varepsilon} \right)$$

corresponds to the minimizer of $\mathcal{F}_{\varepsilon,x}$ under the following boundary conditions:

$$v_\varepsilon(\alpha_\varepsilon) = 0; \ \lim_{s\to\infty} v_\varepsilon(s) = 1.$$

Indeed, it is also possible to construct the field v_ε for the upper Γ-limit along lines intersecting the crack set at 90^0 angles, using the solution to the optimal profile problem on each of those. Integration of the result along the crack set will also permit one to recover the upper Γ-limit.

The Γ–convergence result above can be extended to the restriction $\mathcal{F}_{\varepsilon,h}$ of $\mathcal{F}_\varepsilon$ to a linear finite element approximation, provided that the discretization parameter h is such that $h = o(\varepsilon)$ (see (Bellettini and Coscia, 1994), (Bourdin, 1999)). A closer look at the construction for the upper Γ–limit and at its adaptation to $\mathcal{F}_{\varepsilon,h}$ provides some useful insight into possible error estimates.

The construction of the sequence $(\varphi_{\varepsilon,h}, v_{\varepsilon,h})$ for the upper Γ–limit for $\mathcal{F}_{\varepsilon,h}$ can be obtained from that above. Let $\mathcal{T}_h$ be a conforming mesh of $\tilde{\Omega}\backslash\partial\Omega_s$ and S_h be the set of all elements in $\mathcal{T}_h$ intersecting $S(\varphi)$. Let π_h be a linear finite element projection operator associated with $\mathcal{T}_h$, and consider

$$v_{\varepsilon,h}(x) := \begin{cases} 0 & \text{if } x \in S_h; \\ \pi_h(v_\varepsilon) & \text{otherwise,} \end{cases} \qquad (8.10)$$

and

$$\varphi_{\varepsilon,h}(x) := \pi_h(\varphi_\varepsilon). \qquad (8.11)$$

Following a path similar to that developed in the computation of the upper Γ-limit above, the first term $\ell(\alpha_\varepsilon)/4\varepsilon$ on the right hand-side of inequality (8.7) becomes $|S_h|/4\varepsilon \simeq \mathcal{H}^1(S(\varphi))h/4\varepsilon$, which converges to 0 only if $h = o(\varepsilon)$. The consideration of quadratic finite elements in lieu of linear ones would still induce an error on the surface energy of the order of h/ε, albeit with a different constant. This is why the proposed implementation only resorts to piecewise linear finite elements for φ and v.

In a different direction, this term links the anisotropy of the mesh to the quality of the approximation of the surface energy. In (Negri, 1999), M. Negri studied the effect of various types of structured meshes on the surface energy for the Mumford-Shah problem. In the numerical experiments, the isotropy of the surface term is ensured through the use of "almost" isotropic Delaunay meshes.

From the construction above, it is deduced that the relation $h = o(\varepsilon)$ only needs to be satisfied "close" to $S(\varphi)$. Of course, barring prior knowledge of $S(\varphi)$, uniformly homogeneous fine meshes are a must.

However, *a posteriori* re-meshing the domain will then improve the accuracy of the energy estimate. Note that *a priori* mesh adaption – or setting mesh adaptation as an integral part of a minimization algorithm – can prove slippery. Because the local size of the mesh affects the quality of the approximation of the surface energy, such a process could potentially create spurious local minimizers. So, *a posteriori* mesh refinement around the cracks shields the computations from artificial cracks that would correspond to local minima created by *a priori* mesh refinement!

The sequence for the upper Γ–limit is also admissible for the lower Γ–limit, so that, if φ if a minimizer for the total energy, the sequence $(\varphi_{\varepsilon,h}, v_{\varepsilon,h})$ constructed above approximates a minimizing sequence for $\mathcal{F}_{\varepsilon,h}$ and this asymptotically in h, that is in particular

$$ k \int_{\widetilde{\Omega} \backslash \partial \Omega_s} \left\{ \frac{(1 - v_{\varepsilon,h})^2}{4\varepsilon} + \varepsilon |\nabla v_{\varepsilon,h}|^2 \right\} dx \cong k \left(1 + \frac{h}{4\varepsilon} \right) \mathcal{H}^1(S(\varphi)). $$

$$(8.12)$$

In practice, it is as if the fracture toughness k had been amplified by a factor $1 + h/(4\varepsilon)$, yielding an effective toughness $k_{eff} = k(1 + h/(4\varepsilon))$ which has to be accounted for when interpreting the results. The experiments in Section 8.3.2 highlight the effect of mesh isotropy on the results, and show how the fracture toughness is overestimated.

8.1.2. *Quasi-static evolution*

The approximation scheme devised in Subsection 8.1 should now be reconciled with the evolutionary character of the weak discrete formulation. Irreversibility of the crack growth is enforced at the time-discrete level in the manner described below.

Consider a fixed ε and a fixed conforming mesh $\mathcal{T}_h$ of $\widetilde{\Omega} \backslash \partial \Omega_s$ with characteristic element size h. Introduce a small parameter $\eta > 0$, and at each step t_i, the set of vertices

$$ K_{\varepsilon,h,\eta}^i := \left\{ s \in \mathcal{T}_h \, ; \, v_{\varepsilon,h}^i(s) \le \eta \right\}, \, i > 0; \, K_{\varepsilon,h,\eta}^0 := \emptyset. $$

In the light of the Γ-convergence properties of $\mathcal{F}_{\varepsilon,h}$, the crack growth condition translates into a growth condition on the sets $K_{\varepsilon,h,\eta}^i$ and leads to the following fully spatially and temporally discrete evolution scheme:

(Fde) Find a sequence $\left(\varphi_{\varepsilon,h}^{i+1}, v_{\varepsilon,h}^{i+1} \right)_{i=0,\ldots,n}$ of global minimizers for $\mathcal{F}_{\varepsilon,h}$ under the constraints

$$ \varphi = g(t_{i+1}) \text{ on } \widetilde{\Omega} \backslash \Omega $$

and
$$v = 0 \text{ on } K^i_{\varepsilon,h,\eta}. \tag{8.13}$$

Recently, Giacomini conducted a rigorous analysis of a slightly different approach to the time evolution for $\mathcal{F}_\varepsilon$. In (Giacomini, 2005a), crack growth is enforced through the monotonicity of v in time, *i.e.*, by successively minimizing $\mathcal{F}_\varepsilon$ among all (φ, v) such that $\varphi = g(t_{i+1})$ on $\tilde{\Omega} \setminus \Omega$, and $v \le v_i^\varepsilon$ almost everywhere on Ω. In that setting, as both the time discretization parameter (Δ_n) and ε go to 0 (in a carefully ordered fashion), the discrete evolution converges to a continuous evolution satisfying the conclusions of Theorem 5.5.

In the forthcoming numerical experiments, the monotonicity constraint is imposed as described in (8.13). Implementing Giacomini's constraint in its place would not generate additional difficulties, but would increase the computational cost.

REMARK 8.3. The Γ-convergence based approach to minimization is not so easily amenable to the treatment of local minimization. If $(\varphi, 1)$ is an *isolated* L^1-local minimizer for $\mathcal{F}$ (see (8.1)), then Theorem 2.1 in (Kohn and Sternberg, 1989) can be adapted to the current setting to prove the existence of a sequence of L^1-local minimizers $(\varphi_\varepsilon, v_\varepsilon)$ for $\mathcal{F}_\varepsilon$ converging to $(\varphi, 1)$ in L^1. Unfortunately, the isolation hypothesis is generically false: see for instance the 1d-traction experiment with a hard device in Paragraph 3.1.2.

Even when the isolation hypothesis applies, the above-mentioned theorem grants the existence of a sequence of local minimizers for $\mathcal{F}_\varepsilon$ converging to a local minimizer of $\mathcal{F}$, but does not however guarantee that a converging sequence of local minimizers for $\mathcal{F}_\varepsilon$ converges to a local minimizer for $\mathcal{F}$.

For further considerations on the convergence of local minimizers and/or stationary points, see (Francfort et al., 2008).

8.2. MINIMIZATION ALGORITHM

Recall that $\mathcal{F}_{\varepsilon,h}$ is the restriction of $\mathcal{F}_\varepsilon$ defined in (8.2) to a linear finite element approximation. Also note that, although $\mathcal{F}_\varepsilon$ is separately convex in its arguments φ and v, it is not convex in the pair (φ, v).

In the numerical experiments below, we fix the regularization parameter ε and generate a mesh with characteristic size h. We do not try to adapt the values of ε and h during the numerical minimization of $\mathcal{F}_{\varepsilon,h}$. Thus, the numerical implementation reduces to a sequence of minimizations for $\mathcal{F}_{\varepsilon,h}$, each corresponding to a separate time step. All presented experiments have been tested on meshes of various size and

with different values of the parameter ε and/or of the time discretization length; the results seem impervious to such changes, at least for
reasonably small choices of the parameters ε, h, Δ_n.

Unfortunately, contrary to the adage, you can tell a functional by
its cover and the lack of convexity of $\mathcal{F}_{\varepsilon,h}$, inherited from that of $\mathcal{F}_\varepsilon$,
promptly dashes any hope for a fool-proof minimization scheme. As
per Section 8.1, we should choose a mesh size h which remains "small"
compared to the regularization parameter, which in turn needs to be
"small". In a 2d setting, this typically results in meshes with $(10)^5$
elements, while in three dimensions, the mesh used in the experiment
shown on Figure B, page 101, consists of over $1.7\ (10)^6$ elements.
Although the analysis of such large problems can be tackled thanks
to the wider availability of massively parallel computers, there are, to
our knowledge, no global minimization algorithms capable of handling
them. At best, the algorithms will asymptotically satisfy necessary
optimality conditions for minimality.

8.2.1. *The alternate minimizations algorithm*

The first building block in the numerical implementation is an alternate
minimization algorithm, leading to evolutions satisfying a first set of
necessary conditions for optimality.

The functional $\mathcal{F}_\varepsilon$ – and therefore $\mathcal{F}_{\varepsilon,h}$ – is Gâteaux-differentiable
around any (φ, v). We compute the first order variation of $\mathcal{F}_{\varepsilon,h}$ around
any kinematically admissible (φ, v) in the directions $(\tilde{\varphi}, 0)$ and $(0, \tilde{v})$,
where $\tilde{\varphi}$ and $\tilde{v}$ are admissible variations ($\tilde{\varphi} = 0$ on $\widetilde{\Omega} \backslash \Omega$ and $\tilde{v} = 0$ on
$K^i_{\varepsilon,h,\eta}$) and obtain that the solution $(\varphi^{i+1}_{\varepsilon,h}, v^{i+1}_{\varepsilon,h})$ of the fully discrete
evolution at time step t_{i+1} satisfies

$$
\begin{cases}
\displaystyle \int_\Omega \left((v^{i+1}_{\varepsilon,h})^2 + \eta_\varepsilon \right) DW(\nabla \varphi^{i+1}_{\varepsilon,h}) \cdot \nabla \tilde{\varphi}\, dx = 0 \\[4mm]
\displaystyle \int_\Omega \left(v^{i+1}_{\varepsilon,h} \tilde{v} \right) W(\nabla \varphi^{i+1}_{\varepsilon,h})\, dx + k \int_{\widetilde{\Omega} \backslash \partial \Omega_s} \left\{ \frac{v^{i+1}_{\varepsilon,h} \tilde{v}}{4\varepsilon} + \varepsilon \nabla v^{i+1}_{\varepsilon,h} \cdot \nabla \tilde{v} \right\} dx \\[4mm]
\hspace{5cm} = k \displaystyle \int_{\widetilde{\Omega} \backslash \partial \Omega_s} \frac{\tilde{v}}{4\varepsilon}\, dx.
\end{cases}
\tag{8.14}
$$

This leads to the following algorithm, where δ is a fixed tolerance
parameter:

ALGORITHM 1. *The alternate minimizations algorithm:*
  1: Let $p = 0$ and $v^{(0)} := v^i_{\varepsilon,h}$.
  *2: **repeat***
  3: $p \leftarrow p+1$

4:   *Compute* $\varphi^{(p)} := \arg\min_{\varphi} \mathcal{F}_{\varepsilon,h}(\varphi, v^{(p-1)})$ *under the constraint* $\varphi^{(p)} = g(t_{i+1})$ *on* $\widetilde{\Omega}\backslash\Omega$.

5:   *Compute* $v^{(p)} := \arg\min_{v} \mathcal{F}_{\varepsilon,h}(\varphi^{(p)}, v)$ *under the constraint* $v^{(p)} = 0$ *on* $K^i_{\varepsilon,h,\eta}$

6:   **until** $\|v^{(p)} - v^{(p-1)}\|_{\infty} \leq \delta$

7:   *Set* $\varphi^{i+1}_{\varepsilon,h} := \varphi^{(p)}$ *and* $v^{i+1}_{\varepsilon,h} := v^{(p)}$

Since $\mathcal{F}_{\varepsilon,h}$ is separately convex in each of its arguments, the algorithm constructs at each time step a sequence with decreasing total energy; it is therefore unconditionally convergent in energy. A more detailed analysis conducted in (Bourdin, 2007a) proves that, whenever the cracks are *a priori* known to propagate smoothly, the alternate minimization algorithm converges to the global minimizer of $\mathcal{F}_{\varepsilon,h}$ for fine enough time discretization steps. In cases where cracks propagate brutally, this algorithm can only be proved to converge to critical points of $\mathcal{F}_{\varepsilon}$, which may be a local (or global) minimizers, but also saddle points for $\mathcal{F}_{\varepsilon}$. As per Remark 8.3, local minimizers of $\mathcal{F}_{\varepsilon}$ can sometimes be proved to converge to local minimizers of $\mathcal{F}$. Similar results in the case of saddle points are for now restricted to the 1d setting (Francfort et al., 2008). The detection of saddle points require a detailed stability study. Because of the typical size of the problems, this is a difficult task which has yet to be implemented. In its stead, we shift our focus on the derivation of additional necessary conditions for minimality and propose to devise compatible algorithms.

8.2.2. *The backtracking algorithm*

When cracks propagate brutally, the alternate minimizations algorithm, or any other descent-based algorithm for that matter, cannot be expected to converge to the global minimizer of $\mathcal{F}_{\varepsilon,h}$. Indeed, a numerical method that relies solely on (8.14) will lead to evolutions whose total energy $E(t)$ is not an absolutely continuous (or even continuous) function (see Figure 11 in (Negri, 2003) or Figure 3(b) in (Bourdin et al., 2000)). This is incompatible with the convergence of the time-discretized to the time continuous evolution, culminating in Theorem 5.5. So, since (8.14) is satisfied at each time step, those evolutions have to correspond to local minimizers or saddle points of the regularized energy. Such solutions – spurious from the standpoint of global minimization – can actually be eliminated by enforcing an additional optimality condition.

Consider a monotonically increasing load, as in Section 2.3, and suppose the elastic energy density W to be 2-homogeneous (adapting this argument to p-homogenous W is trivial). If $(\varphi^i_{\varepsilon,h}, v^i_{\varepsilon,h})$ is admissible for a time step t_i, then $\left(t_j/t_i\varphi^i_{\varepsilon,h}, v^i_{\varepsilon,h}\right)$ is admissible for all time steps t_j with $0 \leq j \leq i$, and

$$\mathcal{F}_{\varepsilon,h}\left(\frac{t_j}{t_i}\varphi^i_{\varepsilon,h}, v^i_{\varepsilon,h}\right) = \frac{t_j^2}{t_i^2}\mathcal{F}^b_{\varepsilon,h}(\varphi^i_{\varepsilon,h}, v^i_{\varepsilon,h}) + \mathcal{F}^s_{\varepsilon,h}(v^i_{\varepsilon,h}),$$

$\mathcal{F}^b_{\varepsilon,h}$ and $\mathcal{F}^s_{\varepsilon,h}$ denoting respectively the bulk and surface terms in $\mathcal{F}_{\varepsilon,h}$. But if the sequence $\{(\varphi^i_{\varepsilon,h}, v^i_{\varepsilon,h})\}$ is a solution of the fully discrete evolution, $(\varphi^j_{\varepsilon,h}, v^j_{\varepsilon,h})$ must minimize $\mathcal{F}_{\varepsilon,h}$ among all admissible pairs (φ, v), and in particular, for $0 \leq j \leq i \leq n$,

$$\mathcal{F}^b_{\varepsilon,h}\left(\varphi^j_{\varepsilon,h}, v^j_{\varepsilon,h}\right) + \mathcal{F}^s_{\varepsilon,h}\left(v^j_{\varepsilon,h}\right) \leq \frac{t_j^2}{t_i^2}\mathcal{F}^b_{\varepsilon,h}(\varphi^i_{\varepsilon,h}, v^i_{\varepsilon,h}) + \mathcal{F}^s_{\varepsilon,h}(v^i_{\varepsilon,h}). \quad (8.15)$$

Note that in establishing this condition, we used the *global* minimality of the evolution $\{(\varphi^i_{\varepsilon,h}, v^i_{\varepsilon,h})\}$, so that (8.15) is a necessary condition for global minimality but it is neither necessary nor sufficient for local minimality. Since $t_j \leq t_i$, the total energy, that is $\{\mathcal{F}_{\varepsilon,h}(\varphi^i_{\varepsilon,h}, v^i_{\varepsilon,h})\}$, associated with an evolution satisfying (8.15) is monotonically increasing. In the time continuous limit, any such evolution produces an absolutely continuous total energy, in accordance with Theorem 5.5.

Algorithmically, we check condition (8.15) against all previous time steps t_j, with j varying from 0 to i. If for some t_j, (8.15) is not satisfied, then $(\varphi^j_{\varepsilon,h}, v^j_{\varepsilon,h})$ cannot be the global minimizer for the time step t_j, and $(t_j/t_i\varphi^i_{\varepsilon,h}, v^i_{\varepsilon,h})$ provides an admissible field with a strictly smaller energy at time t_j. In this case, we backtrack to time step t_j, and restart the alternate minimizations process, initializing the field v with $v^i_{\varepsilon,h}$. Because the alternate minimizations algorithm constructs sequences with monotonically decreasing energy (at a given time step), repeated backtracking will converge to a solution such that (8.15) is satisfied for this particular choice of i and j.

The backtracking algorithm can be summarized as follows, with δ a small tolerance:

ALGORITHM 2. *The backtracking algorithm:*

1: $v_0 \leftarrow 1,\ 1 \leftarrow i$

2: **repeat**

3:      Compute $(\varphi^i_{\varepsilon,h}, v^i_{\varepsilon,h})$ using the alternate minimization algorithm initialized with v_0.

4:      Compute the bulk and surface energies $\mathcal{F}^b_{\varepsilon,h}(\varphi^i_{\varepsilon,h}, v^i_{\varepsilon,h})$, $\mathcal{F}^s_{\varepsilon,h}(v^i_{\varepsilon,h})$

5:      **for** $j = 1$ to $i - 1$ **do**

6:    **if** $\mathcal{F}_{\varepsilon,h}(\varphi^j_{\varepsilon,h}, v^j_{\varepsilon,h}) - \left(\frac{t_j}{t_i}\right)^2 \mathcal{F}^b_{\varepsilon,h}(\varphi^i_{\varepsilon,h}, v^i_{\varepsilon,h}) - k\mathcal{F}^s_{\varepsilon,h}(v^i_{\varepsilon,h}) \geq \delta$
       then

7:        $v_0 \leftarrow v^i_{\varepsilon,h}$

8:        $i \leftarrow j$

9:        **return** *to 3:*

10:    **end if**

11:   **end for**

12:   $v_0 \leftarrow v^i_{\varepsilon,h}$

13:   $i \leftarrow i+1$

14: **until** $i = n$

REMARK 8.4. The backtracking algorithm is not refined enough to avoid local minimizers. It merely selects, at each time step, the proper critical point among all previously identified potential solutions. In the following subsection, "large" enough loads force bifurcation and crack creation, but other avenues should certainly be explored. As possible alternative, we mention time refinement, *i.e.*, a first set of computations with a coarse time discretization, then a refinement of the time step, or still topology generation, *i.e.*, an initialization with a v field that would represent many "small" cracks constructed using the optimal profile (8.5).

8.3. NUMERICAL EXPERIMENTS

This subsection describes in detail various numerical experiments: 1d traction, anti-plane shear tearing, and 2d tearing of a plate. The first two experiments mirror and expand on the theoretical predictions of Paragraph 3.1.2 and Subsection 3.2 respectively, while the third example is a theoretical "terra incognita".

Computational convenience, rather than mechanical realism guides the choice of the various mechanical quantities. Even worse, we do not specify the units for those quantities; the reader is at liberty to check the dimensional consistency of the various expressions below.

8.3.1. *The 1D traction (hard device)*

At this point, the confusion sown in the reader's mind by the many detours of the proposed algorithm will undoubtedly make her doubt our numerical predictive ability when it comes to fracture. We now propose to put her mind to rest and, to this end, consider the very simple benchmark example of the 1d traction experiment (hard device) from Section 3.1.2.

A long beam of length L and cross section $\Sigma = 1$ is clamped at $x = 0$, and subject to a displacement load tL at its right extremity $x = L$. With $W(a) \equiv 1/2E(a-1)^2$ in (8.2), the pair $(\varphi_e, v_e^\varepsilon)$ defined as

$$\varphi_e := (1+t)x, \quad v_e^\varepsilon := \frac{k}{k + 2\varepsilon E t^2}$$

is immediately seen to be a critical point for $\mathcal{F}_\varepsilon$ for any t, *i.e.*, a solution to the associated Euler-Lagrange equations

$$\begin{cases} ((v^2 + \eta_\varepsilon)\varphi')' = 0 \\ \dfrac{1}{2}vE(\varphi'-1)^2 - \varepsilon v'' - k\dfrac{(1-v)}{4\varepsilon} = 0 \\ \varphi(t,0) = 0, \ \text{and} \ \varphi(t,L) = (1+t)L. \end{cases}$$

Further,

$$\mathcal{F}_\varepsilon(\varphi_e, v_e^\varepsilon) = \frac{t^2 EkL}{2(k + 2\varepsilon E t^2)},$$

where, for algebraic simplicity, the term η_ε has been dropped from the expression (8.2) for $\mathcal{F}_\varepsilon$. The Γ–limit result of Subsection 8.1 guarantees the existence of a sequence $\{(\varphi_1^\varepsilon, v_1^\varepsilon)\}$ such that

$$\varphi_1^\varepsilon \to \varphi_1 := \begin{cases} x, \ x < x_1 \\ x + tL, \ x > x_1 \end{cases} \quad \text{for some } x_1 \in [0, L]$$

$$v_1^\varepsilon \to 1 \ \text{a.e. in } (0, L),$$

provided that $t > \sqrt{2k/EL}$ (see (3.4) with t replacing ε in that expression). It also guarantees that

$$\mathcal{F}_\varepsilon(\varphi_1^\varepsilon, v_1^\varepsilon) \to k,$$

the energy associated with the solution with one jump.

It is immediate that the alternate minimization initiated with $v \equiv 1$ converges to $(\varphi_e, v_e^\varepsilon)$ in one iteration. The same analysis applies to any problem whose elastic solution has constant gradient. Indeed, it seems preposterous to expect that the proposed numerical method should ever converge towards a solution with cracks!

The solution of this conundrum requires a detailed study of the stability of the critical point $(\varphi_e, v_e^\varepsilon)$; see (Bourdin, 2007a, Section 3.1). It is shown there that, given *any* admissible $\tilde{\varphi}$ such that $\tilde{\varphi} \neq 0$, there exist $\tilde{v} \neq 0$ and $t^\varepsilon(\tilde{\varphi})$ such that

$$\mathcal{F}_\varepsilon(\varphi_e + \alpha\tilde{\varphi}, v_e^\varepsilon + \alpha\tilde{v}) < \mathcal{F}_\varepsilon(\varphi_e, v_e^\varepsilon),$$

when $t \geq t^{\varepsilon}(\tilde{\varphi})$, and for small enough α. In other words, any direction will become a direction of descent for the energy past a direction-dependent critical load t^{ε}, and $(\varphi_e, v_e^{\varepsilon})$ will then become a saddle point. In the context of the alternate minimization algorithm, the discretization error is sufficient to induce the bifurcation of the minimization algorithm away from $(\varphi_e, v_e^{\varepsilon})$. Numerically, this is exactly what is being observed; the critical load at which the numerical solution bifurcates from the elastic to the cracked solution increases when $\varepsilon \to 0$ or $h \to 0$. Once the bifurcation occurs, and the alternate minimizations identify $(\varphi_1^{\varepsilon}, v_1^{\varepsilon})$ as another critical point, the backtracking algorithm leads to the proper detection of the critical load and the reconstructed evolution matches (3.4).

Figure 8.1 follows the evolution of the energy for this experiment. The parameters are $L = 10$, $E = 4\,(10)^{-2}$, $k = 1$, the mesh size is $h = 1.5\,(10)^{-2}$, and $\varepsilon = 8\,(10)^{-2}$. For those parameters, the critical load at which fracture occurs is $t_c = \sqrt{5} \simeq 2.24$, according to (3.4). Without backtracking, the critical load upon which the solution bifurcates from the uncracked to the cracked solution is approximately 7.85, and the total energy is clearly not continuous.

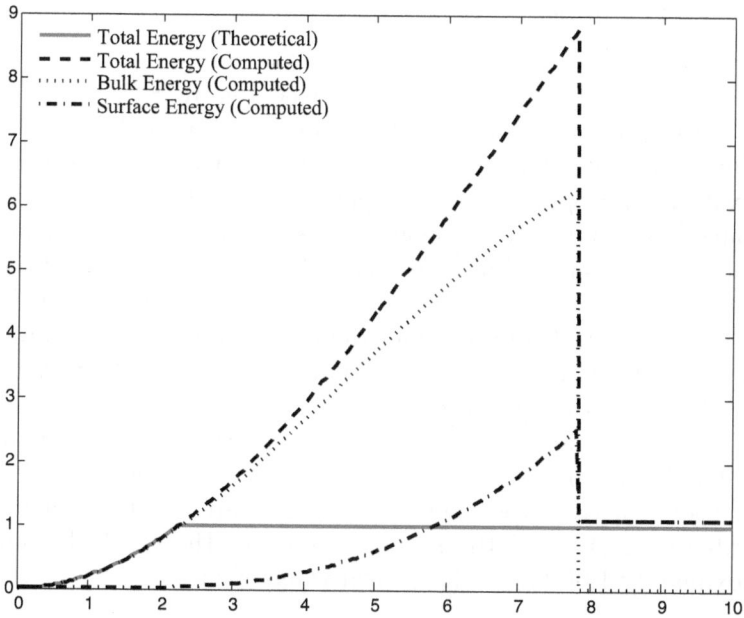

Figure 8.1. Evolution of the total, bulk and surface energies for the 1d traction experiment (hard-device) without backtracking

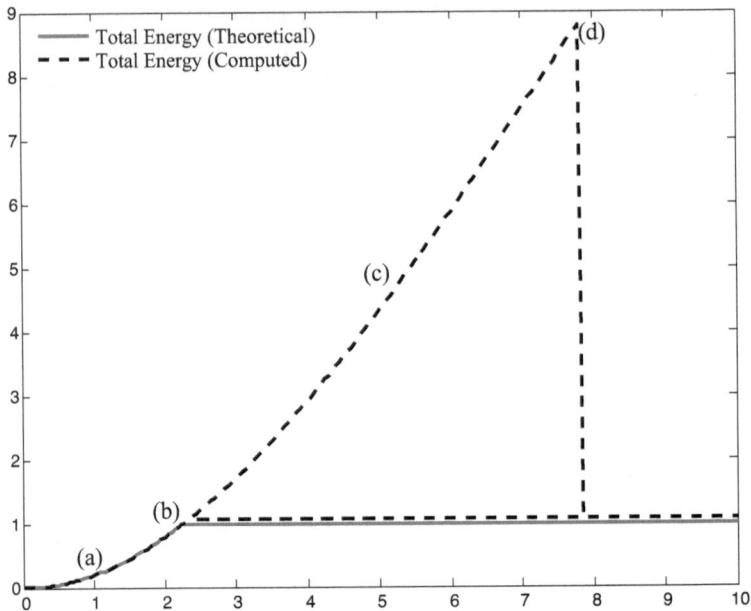

Figure 8.2. Evolution of the total energy for the 1d traction experiment (hard-device) with backtracking

Using the backtracking algorithm (see Figure 8.2) allows one to recover the proper evolution. The reader is reminded of the note of caution offered in the Introduction: the backtracking algorithm is used here with $u(x) := \varphi(x) - x$, the displacement field, as variable, because the bulk energy is 2-homogeneous in $\varphi' - 1$ and not in φ'. At first, the computed solution is similar to that obtained without backtracking. When t reaches 7.85, the alternate minimizations algorithm bifurcates towards the cracked solution, and the total energy decreases (step (d) in Figure 8.2). At that point, the optimality condition (8.15) is violated for all $2.4 \leq t \leq 7.85$ (marked (c)).

The alternate minimization is then restarted from $t = 2.4$ (marked (b)). The final evolution closely matches the theoretical solution. The critical load in the experiments is approximately 2.4 (vs. a theoretical value of 2.24), and the surface energy of the cracked solution is approximately 1.08 (vs. a theoretical value of 1).

8.3.2. *The tearing experiment*

The second numerical simulation follows along the lines of the tearing experiment investigated in Section 3.2. We consider a rectangular domain $\Omega = (0, L) \times (-H, H)$. The analysis in Subsection 3.2 still applies and the field constructed there under assumption (3.14) is an admissible test field for this problem, provided of course that $0 \ll l(t) \leq L$.

However, when the domain has finite length, a crack splitting the whole domain is a minimizing competitor. Let φ_c represent that solution. Following the notation in Section 3.2, we set

$$\begin{cases} S(\varphi_c) = (0, L) \times \{0\} \\ u_c(t, x) = tH, \end{cases}$$

so that

$$E(S(\varphi_c)) = kL.$$

A comparison of the energy of both types of evolutions demonstrates that, under assumption (3.14), the global minimizer for the tearing problem is such that $\mathbf{u}(x, y, t) = \text{sign}(y)u(t, x)\mathbf{e}_3$ and $S(\varphi) = [0, l(t)) \times \{0\}$, with

$$u(x, t) = \begin{cases} tH \left(1 - \dfrac{x}{l(t)}\right)^{+} & \text{if } t \leq \dfrac{L}{2H}\sqrt{\dfrac{k}{\mu H}} \\ tH & \text{otherwise,} \end{cases} \tag{8.16}$$

where

$$l(t) = \begin{cases} tH\sqrt{\dfrac{\mu H}{k}} & \text{if } t \leq \dfrac{L}{2H}\sqrt{\dfrac{k}{\mu H}} \\ L & \text{otherwise.} \end{cases} \tag{8.17}$$

This corresponds to a crack that propagates at constant speed

$$\frac{dl}{dt} = H\sqrt{\frac{\mu H}{k}}$$

along the symmetry axis, until its length reaches $L/2$, and then jumps along the x-axis until the end point of that axis in the domain. Note that, during the smooth propagation phase, the bulk and surface energies of the sample are equal, and that, throughout the evolution, the total energy of the solution is

$$E(t) = \min\left(2tH\sqrt{\mu Hk}, kL\right). \tag{8.18}$$

We wish to illustrate the ability of the advocated numerical approach to capture the proper evolution for a known crack path. As a first step in that direction, the anti-plane tearing problem is numerically solved by a method developed in (Destuynder and Djaoua, 1981), then compared to the crack evolution analytically obtained above.

We consider a domain with dimensions $H = 1$, $L = 5$. The material properties are $E = 1$, $\nu = .2$ (corresponding to $\mu \simeq .4167$), $k = 1.25 \, (10)^{-2}$ (corresponding to the value of k_{eff} in (8.12); in the case where $\varepsilon = h$, the "material" fracture toughness is $k = (10)^{-2}$). Following Subsection 3.2, the analysis is restricted at first to symmetric solutions consisting of a single crack of length $l(t)$ propagating along the x–axis, starting from the left edge of the domain, with $l(0) = 0$. In order to estimate $l(t)$, we compute the equilibrium deformation $\varphi(1, l)$ corresponding to a unit load and a crack of length l, using finite element meshes consisting of approximately 70,000 nodes, automatically refined around the crack tip. For various choices of $l \in [0, L]$, we estimate the elastic energy $E_b(1, l)$ associated with $\varphi(1, l)$, as well as the energy release rate $G(1, l) = -\partial E_b / \partial l(1, l)$, using classical formulae for the derivative of W with respect to the domain shape. Figures 8.3, 8.4 respectively represent the evolution of $E_b(1, l)$ and $G(1, l)$ as a function of l.

From now onward, we refer to the analytical solution as the "1d solution" in all figures, as well as in the text.

A quick analysis of the numerical results shows that $G(1, l)$ is strictly decreasing (and therefore that W is strictly convex) for $0 \le l < l_c^*$, with $l_c^* \simeq 4.19$. For $l_c^* \le l \le 5$, G is an increasing function of l. Following Proposition 2.4 in Section 2, we deduce that the crack will first propagate smoothly, following Griffith's criterion. When it reaches the length l_c^*, it will then jump brutally to the right edge of the domain

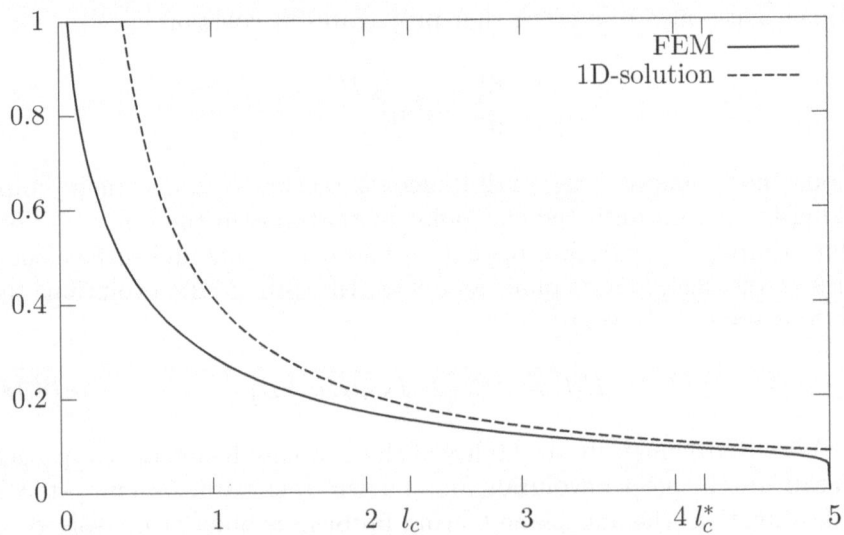

Figure 8.3. Tearing experiment: $l \mapsto W(1, l)$

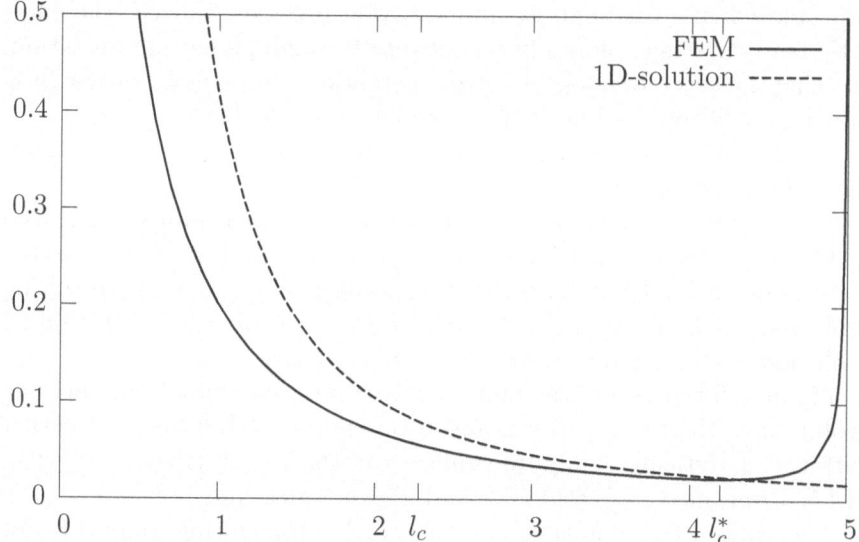

Figure 8.4. Tearing experiment: $l \mapsto G(1, l)$

because not doing so would violate the constraint that $G \leq k$. It could be argued that such an evolution satisfies (necessary conditions for) (Ulm). We will comment further on this evolution in Remark 8.5 and Figure 8.9.

The numerical values of $W(1, l)$ lead to an estimate of the position of the crack tip as a function of the load. Let $\varphi(t, l)$ be the equilibrium deformation associated with the load t, and $E_b(t, l) := t^2 E_b(1, l)$ the associated bulk energy. If the crack keeps on propagating smoothly, then

$$-t^2 \frac{\partial E_b}{\partial l}(1, l) = k. \tag{8.19}$$

That relation is used to compute the load t for which the crack length is l, and thereafter $l(t)$.

Once again, a crack splitting the whole domain along the x–axis is a minimizing competitor. Consider t_c and $l_c := l(t_c)$ such that $E_b(t_c, l_c) + kl_c = kL$. For $t > t_c$, splitting the domain is energetically preferable. The value of t_c can be estimated from the computations of $E_b(1, l)$. Using the finite element computations described above, we get $t_c \simeq .47$. The critical length l_c is such that

$$E_b(1, l_c) = -(L - l_c) \frac{\partial E_b}{\partial l}(1, l_c).$$

Numerically, we obtain $l_c \simeq 2.28$. That value is strictly less than the length l_c^* for which the constraint $G \leq k$ can no longer be met, as expected when global energy minimization presides. Indeed, the energetic landscape is explored in its entirety through global minimization, allowing the crack to adopt a better energetic position at l_c, rather than waiting for G to stumble upon the constraint k at l_c^*.

As an aside, note that the critical length l_c does not depend upon the fracture toughness k!

The attentive reader will have noticed the sudden jump introduced in Griffith's evolution – that satisfying (8.19) – at t_c. Strict orthodoxy would not allow for such a jump to take place, and the resulting evolution might be indicted for revisionism. In all fairness, Griffith's evolution would grind to a halt at l_c^* in any case.

Figure 8.5 represents the numerically computed globally minimizing evolution of the bulk, surface, and total energies (thin lines), together with the analytically computed energies of the 1d solution – see (8.16), (8.17) – obtained in Section 3.2 and above (thick lines).

The computed evolution has the crack propagating smoothly for $0 \le t < t_c$, until it reaches the critical length l_c, then cutting brutally through the domain. For small loads, the one-dimensional analysis

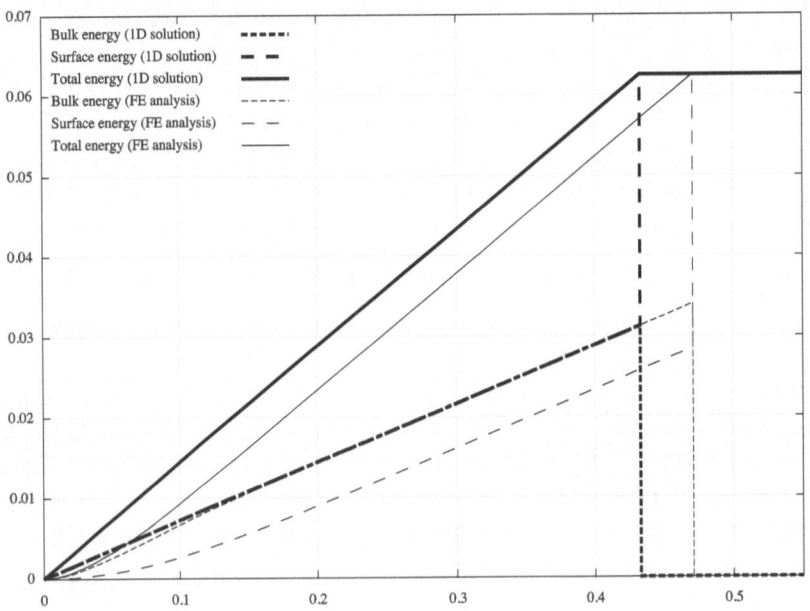

Figure 8.5. Evolution of the bulk surface and total energies following (Ugm), as a function of the load t. They are computed using a classical finite element analysis and compared to the 1d solution

overestimates the crack length; note that as $l \to 0$, $G(1, l) \to \infty$, and that the accuracy of our finite element computations cannot be guaranteed. When t, and therefore l, become large enough, the values of $dE_b(t)/dt$ and $dE_s(t)/dt$ become very close to those obtained in Section 3.2. Numerically we obtain $dE_b(t)/dt \simeq 7.43 \, (10)^{-2}$ and $dE_s(t)/dt \simeq 6.83 \, (10)^{-2}$ while the 1d result is $dE_b(t)/dt = dE_s(t)/dt = H\sqrt{k\mu H} \simeq 7.22 \, (10)^{-2}$.

Next, a numerical experiment that uses the algorithms developed in this section is conducted. We unabashedly reassert our bias towards symmetric solution, resorting to a structured mesh obtained by a split of each square in a structured grid into two right triangles. It consists of 154,450 nodes and 307,298 elements. The mesh size is $(10)^{-2}$; the regularization parameters are $\varepsilon = (10)^{-2}$ and $\eta_\varepsilon = (10)^{-9}$. We consider 100 equi-distributed time steps between 0 and 1. As already noted, the effective toughness in the computations is $k_{\text{eff}} = (1 + h/4\varepsilon)k = .0125$.

Figure 8.6 represents the computed bulk, surface and total energy, as well as their values obtained via the proposed algorithm, as a function of t. Once again, the backtracking algorithm leads to an evolution with a monotonically increasing and continuous total energy.

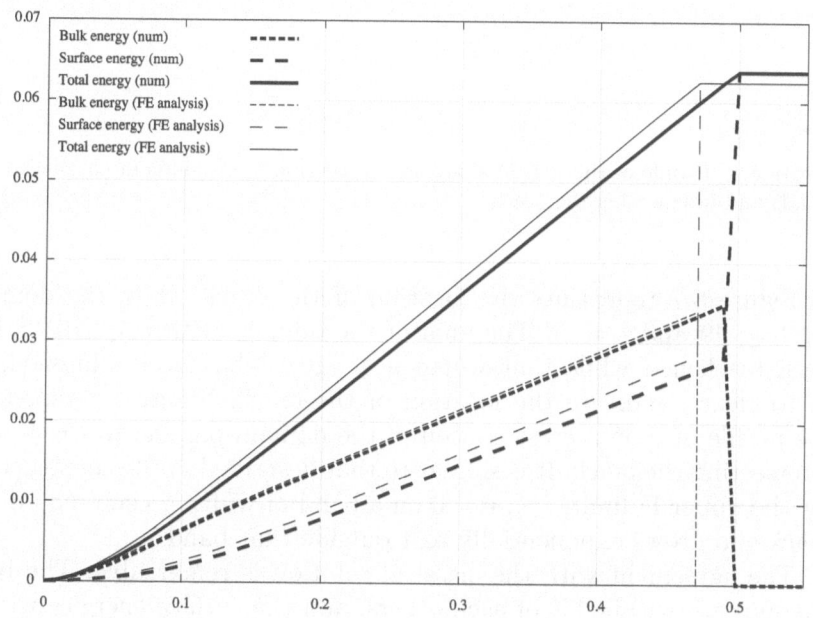

Figure 8.6. Evolution of the bulk surface and total energies following (Ugm), as a function of the load t. Comparison of values obtained through the variational approximation with backtracking and through finite element analysis

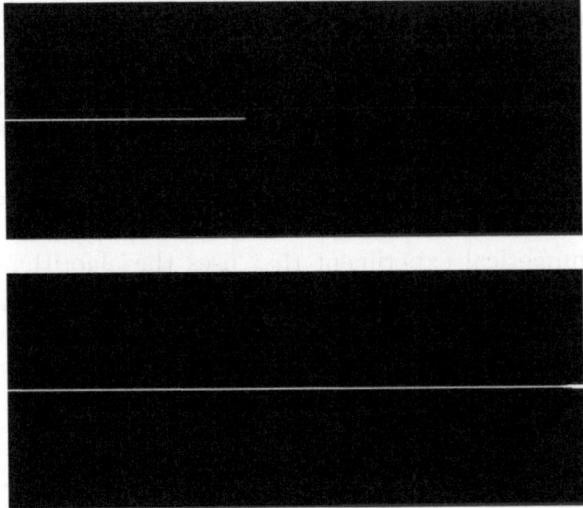

Figure 8.7. Position of the crack set in the tearing experiment for $t = .49$ (*top*) and $t = .50$ (*bottom*)

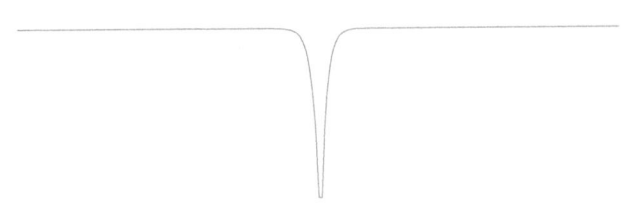

Figure 8.8. Profile of the v–field along a cross section of the domain parallel to the y–axis and intersecting the crack

Figure 8.7 represents the location of the crack set in the domain for $t = .49$ and $t = .5$. The area of the domain with $v \geq (10)^{-1}$ has been blackened while that where $v \leq (10)^{-1}$ has been whitened, so as to clearly indicate the location of the crack. Figure 8.8 represent the profile of v on a cross section of the domain parallel to the y–axis intersecting the crack. It is similar to that described in the construction for the upper Γ–limit, *i.e.*, $v = 0$ on a band of width h centered on the crack and grows exponentially to 1 outside that band.

The agreement with the classical solution is remarkable. The bulk energies are within 1% of each others, and the surface energies within 10%. For long enough cracks, the surface and bulk energies grow at a constant rate, and $dE_b(t)/dt \simeq 6.95\,(10)^{-2}$ and $dE_s(t)/dt \simeq 7.03\,(10)^{-2}$. The critical load upon which the crack propagates brutally is $.49 \leq t_c \leq .5$ (vs. a estimated value of .47), and the critical length is $l_c := l(.49) \simeq$

2.46 which, again, is in agreement with the finite element analysis presented above ($l_c \simeq 2.28$). The final surface energy is $6.38\,(10)^{-2}$, which is consistent with the estimate we gave in Section 8.1.1 ($k(1+h/4\varepsilon)L = 6.25\,(10)^{-2}$).

REMARK 8.5. As noted before, the first evolution computed above using finite element analysis – that is that following Griffith until it jumps at $l_c^* \simeq 4.19$ – can be argued to be one satisfying (necessary conditions for) (Ulm). It propagates smoothly until it reaches $l_c^* \simeq 4.19$ at $t = t_c^* \simeq .75$, then brutally to the right end-side of the domain. Figure 8.9, represent the bulk, surface and total energies of this solution, compared to an experiment using the variational approximation and the alternate minimization, *but without backtracking.* Following the analysis in (Bourdin, 2007a), we expect that, as long as the crack propagates smoothly following local minimizers, the alternate minimization will provide the right evolution. When the crack propagates brutally, nothing can be said. However, once again, the agreement between our experiments is striking. Using the variational approximation, we obtain $t_c^* \simeq .82$ (instead of .75 using the finite element analysis). The estimate for the critical length is $l_c^* \simeq 4.08$ (vs. 4.19 for the finite element computations). Serendipitous or fortuitous?

The symmetry assumption about the x-axis was instrumental in deriving the theoretical results in Subsection 3.2; it was also imposed as a meshing restriction in the previous computation. In its absence, a bona fide theoretical prediction is difficult to make, but an educated guess based on the analogy with e.g. the pre-cracked 2d plate numerically investigated at the onset of Section 6 may provide insight into the possible crack path. We thus introduce a third class of solutions: a crack propagating along the symmetry axis with length $l(t)$ until some critical t_c at which it brutally bifurcates, reaching one of the sides of the domain. The crack for $t \geq t_c$ is assumed L- shaped, *i.e.,* of the form $(0, l(t_c)) \times \{0\} \cup \{l(t_c)\} \times (0, -H)$ or its mirror image with respect to the x–axis. It then remains to minimize in t_c. Appealing to (8.18), (8.17) and comparing the energy associated with the straight crack, *i.e.,* $2tH\sqrt{\mu Hk}$, to that associated with the bifurcated crack, *i.e.,* $k(tH\sqrt{\mu H/k} + H)$, yields

$$t_c = \sqrt{\frac{k}{\mu H}},$$

and

$$l(t_c) = H.$$

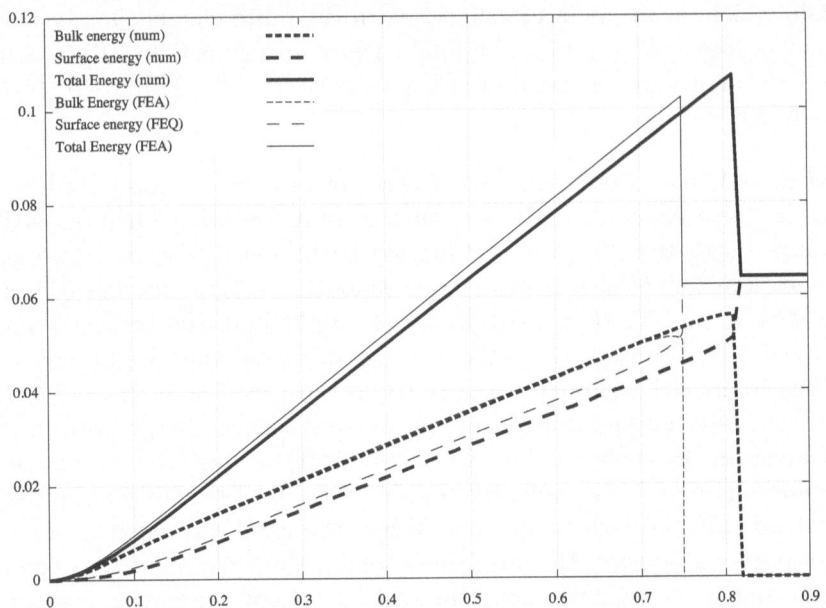

Figure 8.9. Evolution of the bulk surface and total energies following (Ulm), as a function of the load t, computed using a classical finite element analysis. Comparison to the variational approximation without backtracking

The total energy of this branch of solution as a function of the loading parameter t is

$$E(\varphi) = \min\left(2tH\sqrt{\mu Hk}, 2kH\right).$$

If $L > 2H$, this asymmetric solution has a lower energy than its symmetric counterpart as soon as $t \geq \sqrt{k/\mu H}$.

We propose a second set of experiments that use a non-symmetric Delaunay-Voronoi mesh. The mesh size is still $h = (10)^{-2}$, and the other parameters are those of the previous experiment.

The energy plot Figure 8.10 shows that the evolution is qualitatively as expected, *i.e.*, smooth propagation of the crack tip, then brutal propagation.

Once again, the position of the crack tip lags behind its theoretical position and the comparison between the numerical and theoretical energies is difficult.

Figure 8.11 shows the crack tip just before (top) and after (bottom) brutal propagation. The evolution is clearly not globally minimizing: connecting the tip of the crack for $t = .18$ to the upper edge of the domain at a near 90° angle would cost less surface energy. It would be unwise at present to view the perhaps more realistic numerical solution as an outcome of the true minimization. Rather, numerical prudence

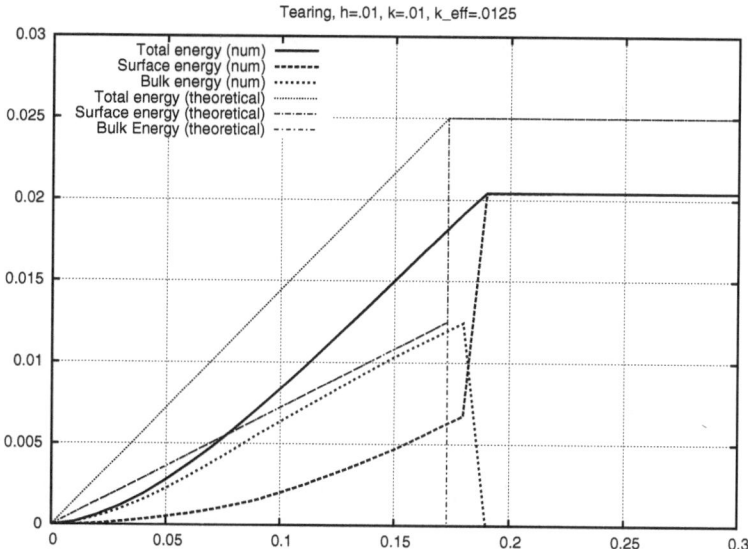

Figure 8.10. Evolution of the bulk surface and total energies as a function of the load t. Numerical and expected values ($t_c \simeq .17$)

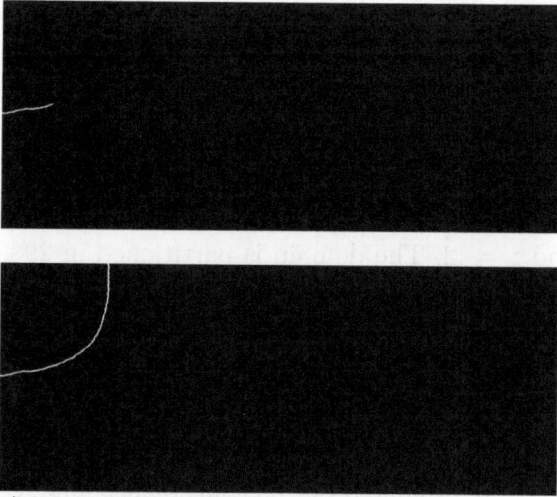

Figure 8.11. Position of the crack set in the tearing experiment for $t = .18$ (*top*) and $t = .19$ (*bottom*)

dictates that it be considered as a lucky bug! This provides a clear illustration of the difficulties of global minimization.

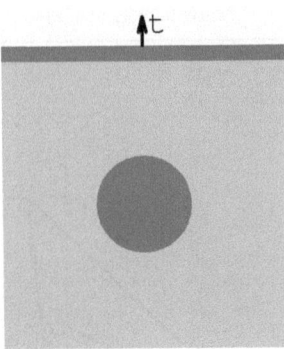

Figure 8.12. 2d traction experiment

8.3.3. *Revisiting the 2D traction experiment on a fiber reinforced matrix*

The numerical experiments above provide ground for a thorough check of the proposed numerical method. However they fall woefully short of target, in that they do not illustrate two of the main tackled issues, initiation and irreversibility in the context of global minimality.

The following example revisits a numerical experiment originally presented in (Bourdin et al., 2000, Section 3.2), and illustrates the improvements brought about by the backtracking algorithm.

A square 2d, brittle and elastic matrix with edge-length 3 is bonded to a rigid circular fiber of diameter 1 as shown in Figure 8.12. The fiber remains fixed, while a uniform displacement field te_2 is imposed on the upper side of the square; the remaining sides are traction free. This is a plane stress problem. The elastic moduli of the matrix are $k = 100$, $E = 4000$, and $\nu = .2$. The domain is partitioned in 293,372 elements and 147,337 nodes, and 125 time steps are used over the interval $0 \leq t \leq .615$. The mesh size is $h = .01$, and the regularization parameters are $\varepsilon = .02$, $\eta_\varepsilon = (10)^{-6}$ (see Paragraph 8.1.1).[14]

The thin lines in Figure C on page 102 show the evolution of the bulk, surface and total energies with respect to t computed without the backtracking algorithm. It is essentially similar to Figure 3-b in (Bourdin et al., 2000). As predicted by the convergence analysis of the alternate minimization algorithm, the total energy is increasing and continuous when the crack propagate smoothly – just correlate the zones where the total energy is continuous with those where the surface energy increases smoothly – but it jumps when the evolution become brutal as witnessed by the total energy restitution associated with the jumps

---

[14] The total computation time is under 2h, using a 32 processors-1.8GHz Xeon cluster

in surface energy at $t = .44$, $t = .47$, and $t = .51$. The critical load at which the alternate minimization bifurcates away from an elastic-type solution is higher than in the experiment in (Bourdin et al., 2000), since the regularization parameter ε is smaller. This is consistent with the stability analysis alluded to in Section 8.3.1.

The thick lines in Figure C on page 102 show the outcome of the same computation, using the backtracking algorithm. The violations of the optimality condition (8.15) were successfully detected and the post-bifurcation solutions used as starting point when restarting the alternate minimization algorithm. The resulting evolution is monotonically increasing and continuous, as predicted by the theory. It is described as follows and shown on Figure D, page 102.

- For $t < .28$, the matrix remains purely elastic, the v field remains close to 1 on the entire domain, and the total energy is a quadratic function of the time;

- At $t \simeq .28$, a curved crack of finite length brutally appears near the top of the inclusion. The increase of the surface energies at that load is exactly balanced by the decrease of bulk energy. The brutal onset of the cracking process agrees with the result obtained in Proposition 4.3 because the crack appears at a non-singular point, thus the initiation time must be positive and the onset brutal;

- For $.28 < t < .38$, the crack grows progressively. The surface energy increases smoothly, while the bulk energy is nearly constant. The propagation is symmetric;

- At $t \simeq .38$, the right ligament breaks brutally, and once again the total energy is conserved. Despite the symmetry of the problem, we obtain an asymmetric solution, which is consistent with the lack of uniqueness of the solution for the variational formulation. Of course, the configuration corresponding to a mirror symmetry of Figure D(d) is also a solution for this time step. That the numerical experiment should favor one solution over the other is purely numerical, and it depends on several factors, including mesh effects – the symmetry of the mesh was not enforced – or rounding errors;

- For $.38 < t < .40$, the remaining crack does not grow, or its propagation is too slow to be detected in the computation. The body stores bulk energy;

- Finally, at $t \simeq .40$, the remaining bulk energy accumulated in the body is released, and the remaining ligament breaks brutally. The domain is split into two parts, and no further evolution takes place.

This numerical experiment, which exemplifies various growth patholo-
gies, compares favorably with an actual experiment reported in (Hull,
1981). Of course the experimental nod of approval is just that, because
the observed agreement is qualitative; a quantitative comparison would
require the design and execution of a carefully tailored experiment, a
task which far exceeds our abilities.

9. Fatigue

Engineering etiquette dictates that a Paris' type phenomenology replace Griffith's model whenever "long" time crack propagation is con-
templated; see (Paris et al., 1961). The substitution remains unmo-
tivated in the literature, with the exception of a few numerical experiments in the cohesive framework, as in (Nguyen et al., 2001) or
in (Roe and Siegmund, 2002). The Paris' type models are difficult to
calibrate and the apportionment of the relevant quantities among such
contributing factors as material properties, geometry and loads is at
best a perilous exercise.

In contrast, we propose to derive Paris' type fatigue laws as a time
asymptotics of the variational model. The three necessary ingredients
are by now familiar to all surviving readers: a minimality principle, a
cohesive type surface energy and irreversibility. The argument is most
easily illustrated on a one-dimensional peeling test; the proofs of all
statements in this section can be found in great details in (Jaubert,
2006), (Jaubert and Marigo, 2006). More general settings could be
envisioned at the expense of mathematical rigor.

An inextensible and flexible thin film with unit width and semi-
infinite length is perfectly bonded at initial time to a rigid substrate
with normal vector $\mathbf{e_2}$. A constant tension $-N\mathbf{e_1}, N > 0$, and a deflection $V_t \mathbf{e_2}$ are applied at its left end ($x = 0$); see Figure 9.1.

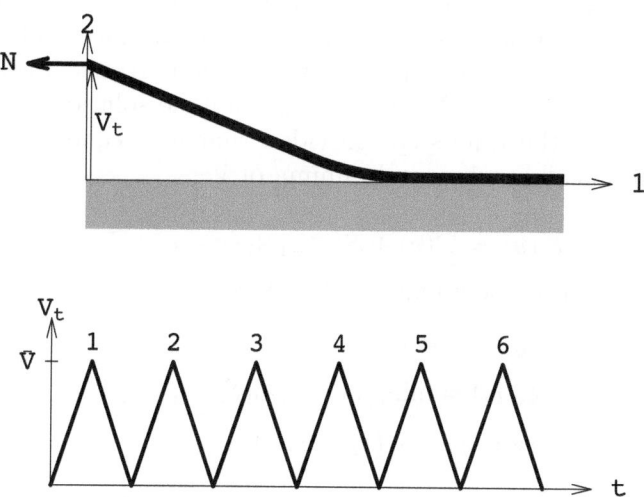

Figure 9.1. Geometry and loading

The displacement of each point x at t is denoted by $\mathbf{U}_t(x) = u_t(x)\mathbf{e_1} + v_t(x)\mathbf{e_2}$; $u_t(\infty) = v_t(\infty) = 0$ and $v_t(0) = V_t$. The deflection V_t periodically oscillates between 0 and $\overline{V}$. The potential energy of the film reduces to the tensile work of N and can be expressed solely in terms of v_t. In a geometrically linear setting, it is of the form

$$\mathcal{P}(v_t) = \frac{N}{2} \int_0^\infty v_t'(x)^2 dx. \qquad (9.1)$$

Debonding is assumed irreversible, so that, in the spirit of Subsection 5.2, we introduce the *cumulated opening* ψ_t as memory variable (see (5.11)), that is here

$$\psi_t(x) = \int_0^t (\dot{v}_\tau(x))^+ d\tau. \qquad (9.2)$$

The perfectly bonded part of the interface at time t corresponds to those points where $\psi_t = 0$, or still where $v_\tau = 0, \forall \tau \le t$.

The selected surface energy density κ_d is that of Dugdale, namely $\kappa_d(\psi) = \min\{\sigma_c\psi; k\}$. The resulting surface energy is thus

$$\mathcal{S}(\psi_t) = \int_0^\infty \kappa_d(\psi_t(x))dx.$$

As hinted at in Subsection 5.2, it is always simpler to investigate the incremental evolution. Although passing to the time-continuous evolution is generally a non-trivial task, as illustrated in the case of Griffith in Subsection 5.1, it can be carried out in the present one-dimensional setting; the interested reader is referred to (Ferriero, 2007). For the loading at hand, that is for a periodic displacement load V_t as represented on Figure 9.1, it is shown in (Jaubert and Marigo, 2006) that the incremental evolution admits a unique solution, and this independently of the time step, provided that the sequence of discrete times contains all maxima and minima of V_t.

With
$$\mathcal{E}_i(v) := \mathcal{P}(v) + \mathcal{S}(\psi_{i-1} + (v - v_{i-1})^+),$$
the incremental problem may be stated as

$$v_0 = \psi_0 = 0$$
$$\mathcal{E}_i(v_i) = \min_{v; v(0)=0, v \ge 0} \mathcal{E}_i(v), \ i \ge 1$$
$$\psi_i = \psi_{i-1} + (v_i - v_{i-1})^+.$$

9.1. PEELING EVOLUTION

As demonstrated in (Jaubert and Marigo, 2006), the cumulated opening ψ remains unchanged during the unloading part of each cycle, while the opening v actually cancels at the bottom of the unloading phase. Consequently, the analysis focusses on the loading part of each cycle and the index i will henceforth refer to the top point of the loading phase of each cycle.

The incremental problem above is easily solved in the case of Griffith's surface energy, that is whenever

$$\kappa_0(\psi) := \begin{cases} k, & \text{if } \psi \neq 0 \\ 0, & \text{if } \psi = 0 \end{cases}$$

is used in lieu of κ_d. Specifically, during the first loading phase, the debond length l grows according to

$$l(t) = \eta V_t,$$

where

$$\eta := \sqrt{\frac{N}{2k}}.$$

Then, it stops at

$$l_1^0 = \eta \overline{V}, \tag{9.3}$$

that is at the top of the first loading phase, not to ever grow again during the subsequent loading phases.

There is "no hope without trouble, no success without fatigue"[15], and Griffith's model well publicized failure is unequivocal in spite of our modest import of the cumulated opening as memory variable.

In the case of Dugdale's model, the cohesive force at x vanishes once the cumulated opening $\psi_i(x)$ is greater than the critical value

$$d := \frac{k}{\sigma_c}.$$

Since the field $x \mapsto \psi_i(x)$ is decreasing (see (Jaubert and Marigo, 2006), Proposition 4), three zones are present at the end of the ith loading half-cycle. Those are (see Figure 9.2)

1. The perfectly bonded zone, that is the interval $(\lambda_i, +\infty)$ where the cumulated opening field ψ_i vanishes;

2. The partially debonded zone, also known as process zone, that is the interval (l_i, λ_i) where the cumulated opening field ψ_i takes its values in the interval $(0, d)$;

---

[15] Xavier Marmier – Récits Américains

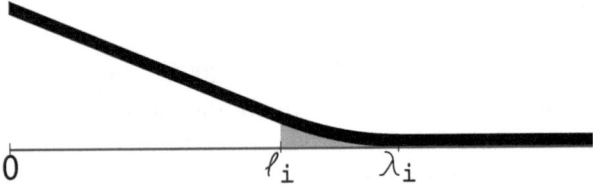

Figure 9.2. The three zones in the case of the Dugdale model

3. The completely debonded zone, also known as the non cohesive zone, that is the interval $(0, l_i)$ where the cumulated opening field ψ_i is greater than d.

Since the cumulated opening at the end $x = 0$ is $\psi_i(0) = i\overline{V}$, the completely debonded zone will eventually appear as the load keeps cycling. For simplicity, we assume that $\overline{V} > d$, which implies the presence of a completely debonded zone at the end of the first half-cycle. At the end of the ith loading phase, the opening field v_i, the cumulated opening field ψ_i and the tips λ_i, l_i of the process and debonded zones are given through the following system of equations

$$N v_i'' = \begin{cases} 0 & \text{in } (0, l_i) \\ \sigma_c & \text{in } (l_i, \lambda_i) \end{cases} \tag{9.4}$$

$$v_i(0) = \overline{V}, \qquad [v_i](l_i) = [v_i'](l_i) = 0, \qquad v_i(\lambda_i) = v_i'(\lambda_i) = 0 \tag{9.5}$$

$$\psi_i(l_i) = \sum_{j=1}^{i} v_j(l_i) = d. \tag{9.6}$$

Equation (9.4) is the Euler equation in the process and debonded zones; the first three equations in (9.5) translate the boundary condition at the end of the film and the continuity conditions at the tips of the zones, while the fourth one is an optimality condition on the position λ_i; (9.6) ensures that the cumulated opening is equal to the critical value d at the tip of the debonded zone and can equally be viewed as an an optimality condition on the position l_i. By virtue of (9.4), (9.5),

$$v_i(x) = \begin{cases} \dfrac{(\lambda_i - l_i)^2}{4\eta^2 d} + \dfrac{(\lambda_i - l_i)}{2\eta^2 d}(l_i - x) & \text{if } 0 \le x \le l_i; \\[2ex] \dfrac{(\lambda_i - x)^2}{4\eta^2 d} & \text{if } l_i \le x \le \lambda_i. \end{cases} \tag{9.7}$$

Inserting (9.7) into (9.5) and (9.6) yields in turn

$$(\lambda_i - l_i)^2 + 2(\lambda_i - l_i)l_i = 4\eta^2 \overline{V} d, \tag{9.8}$$

$$\sum_{j=1}^{i} (\lambda_j - l_i)^{+2} = 4\eta^2 d^2. \tag{9.9}$$

From (9.8) we get

$$\lambda_i = \sqrt{l_i^2 + 4\eta^2 \overline{V} d}$$

while (9.9) implies that the tip l_i of the debonded zone at the ith cycle depends on all previous cycles j which are such that the corresponding tip of the process zone λ_j lies inside the process zone of the ith cycle. The number of such cycles depends on the different parameters η, $\overline{V}$, d and it evolves with the number of cycles. Consequently, (9.9) is a genuinely nonlinear equation for l_i which can only be solved through numerical methods; see Figure 9.3. The sequences $i \mapsto l_i$ and $i \mapsto \lambda_i$ are increasing and "ordered" in that

$$l_{i-1} < l_i < \lambda_{i-1} < \lambda_i, \quad \forall i \geq 2.$$

From this, the onset of fatigue is established in (Jaubert and Marigo, 2006). Specifically,

PROPOSITION 9.1. *For any value $\overline{V} > 0$ of the cycle amplitude, the debond length l_i grows to ∞, the potential energy $\mathcal{P}_i = \mathcal{P}(v_i)$ decreases to 0 and the surface energy $\mathcal{S}_i = \mathcal{S}(\psi_i)$ grows to ∞ as the number of cycles i tends to ∞.*

A cohesive energy and an appropriate memory variable are the key ingredients in producing fatigue. Yet the traditional models of fatigue do not appeal to any kind of yield stress, so that cohesiveness should be flushed out of the model. This is what we propose to achieve in the next subsection. Inspiration will be drawn from Section 7, whose main feature was to view Griffith's model as a limit of cohesive models for very large yield stresses, or still, in the notation of this section, for $d \searrow 0$.

9.2. The limit fatigue law when $d \searrow 0$

As $d \searrow 0$, we assume that the two remaining parameters of the problem, η and $\overline{V}$, are set to fixed values. The tips of the debonded zone and of the process zone at the end of the ith loading phase are now

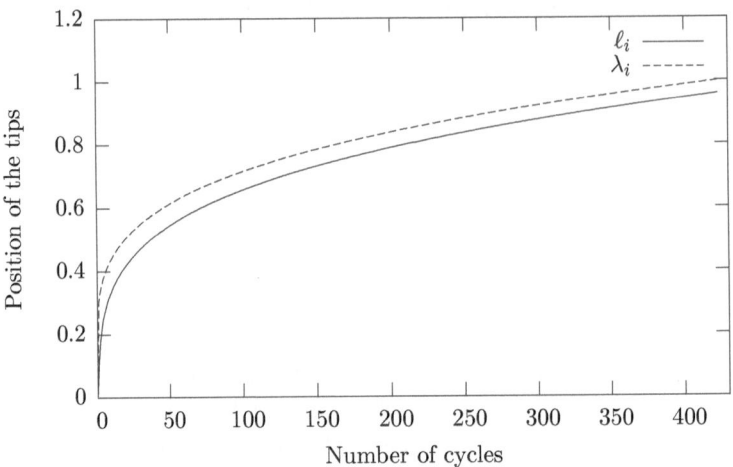

Figure 9.3. Evolution of the tips of the process and debonded zones for $\overline{V} = 0.2$, $d = 0.1$ and $\eta = 1$

denoted by l_i^d and λ_i^d respectively. At the end of the first half-cycle, the position of the tips are given by

$$l_1^d = \eta(\overline{V} - d), \qquad \lambda_1^d = \eta(\overline{V} + d). \qquad (9.10)$$

Note that, when $d = 0$, the result of the Griffith model, that is (9.3), is recovered. For a fixed number of cycles, Dugdale's model "converges" to Griffith's model with d. Indeed (see (Jaubert and Marigo, 2006)),

PROPOSITION 9.2. *For a given number of cycles $i \geq 1$ and when $d \to 0$, the tips λ_i^d and l_i^d tend to l_1^0, i.e., to the debond length given by the Griffith model. Moreover the opening field v_i^d, the potential energy $\mathcal{P}_i^d$ and the surface energy $\mathcal{S}_i^d$ tend to their Griffith analogues at the end of the first half-cycle. In other words,*

$$\lim_{d \to 0} l_i^d = \lim_{d \to 0} \lambda_i^d = \eta\overline{V},$$

$$\lim_{d \to 0} v_i^d(x) = \left(\overline{V} - \frac{x}{\eta}\right)^+, \quad \lim_{d \to 0} \mathcal{P}_i^d = \eta k\overline{V}, \quad \lim_{d \to 0} \mathcal{S}_i^d = \eta k\overline{V}.$$

REMARK 9.3. The reader will readily concede that the above result – which, by the way, agrees in the specific context at hand with Giacomini's cohesive to Griffith analysis of Section 7 – does not contradict Proposition 9.1. Indeed, in that proposition, d is set and the number of cycles goes to infinity, while here the number of cycles is set and d

goes to 0. Straightforward estimates would show that the tip growth at each cycle is of the order of d. Thus, at the ith cycle, the tips are at a position which only differs from that of the first cycle by a distance of the order of d and that difference tends to 0 when d goes to 0.

Consequently, fatigue is a second order phenomenon with respect to the small parameter d. Any hope for fatigue in the non-cohesive limit hinges on a rescaling of the number of cycles of the order of $1/d$, which is precisely what is attempted below.

The number of cycles necessary to debond the film along a given length L is of the order of L/d. Consequently we introduce the positive real parameter T and define the number of cycles $i_d(T)$ by the relation

$$T \longmapsto i_d(T) = \left[\frac{T}{d} \right] \tag{9.11}$$

where $[\cdot]$ denotes the integer part. (Note that T has the dimension of a length.) We also consider numbers of cycles like $i_d(T) + k$, with $k \in \mathbb{Z}$ independent of d.

Figure 9.4 represents the debonded zone tip $l^d_{i_d(T)}$ versus T for different values of d. A Newton-Raphson method is used to compute the solution to (9.9). The graph of $T \mapsto l^d_{i_d(T)}$ is seen to converge to a limit curve $l(T)$ when $d \to 0$. The analytical identification of that curve is the main goal of the remainder of this section.

To that end, we fix $T > 0$ and analyze the asymptotic behavior of the solution at the true number of cycles $i_d(T)$ when d goes to 0. Then (see (Jaubert and Marigo, 2006)),

PROPOSITION 9.4. *At $T > 0$ fixed, when $d \to 0$, the tips $\lambda^d_{i_d(T)}$ and $l^d_{i_d(T)}$ tend to the same limit $l(T)$. Moreover, the opening field $v_{i_d(T)}$, the potential energy $\mathcal{P}_{i_d(T)}$ and the surface energy $\mathcal{S}_{i_d(T)}$ tend to their Griffith analogues, that is*

$$\lim_{d \to 0} l^d_{i_d(T)} = \lim_{d \to 0} \lambda^d_{i_d(T)} = l(T) \geq l^0_1,$$

$$\lim_{d \to 0} v_{i_d(T)}(x) = \left(1 - \frac{x}{l(T)} \right)^+ \overline{V},$$

$$\lim_{d \to 0} \mathcal{P}_{i_d(T)} = \frac{k\eta^2 \overline{V}^2}{l(T)}, \qquad \lim_{d \to 0} \mathcal{S}_{i_d(T)} = kl(T)$$

The process zone is energetically negligible at first order. In other words, the debonding state at T is that of a non cohesive crack of length $l(T)$ with potential energy $\mathcal{P}(T)$ and energy release rate $G(T)$ given by

$$\mathcal{P}(T) = \eta^2 \frac{\overline{V}^2}{l(T)} k, \qquad G(T) = \eta^2 \frac{\overline{V}^2}{l^2(T)} k. \tag{9.12}$$

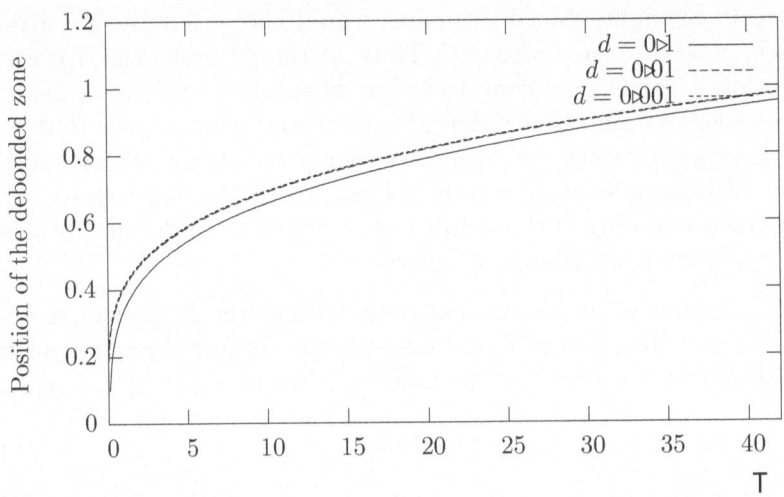

Figure 9.4. Numerical verification of the convergence to a limit curve when $d \to 0$ for $\overline{V} = 0.2$, $\eta = 1$

REMARK 9.5. In view of (9.12), the asymptotic behavior of the film "has the color of [Griffith], tastes like [Griffith], yet it is not [Griffith]"[16] because the value of the debond length $l(T)$ is not that predicted by Griffith's model! Indeed, Griffith's criterion would require $G(T) = k$ and hence that the debond length be $l_1^0 = \eta \overline{V}$, which it is clearly not in view of Figure 9.4.

Also, the function $l(T)$ is monotonically increasing in T. Thus, it admits a limit as $T \searrow 0^+$. Since, by virtue of Proposition 9.2, for fixed d, $\lim_{T \searrow 0^+} l_{i_d(T)}^d = l_1^0 = \eta \overline{V}$, a diagonalization argument would show that $\lim_{T \searrow 0^+} l(T) = l_1^0$.

We assume that the macroscopic debond length $l = l(T) > l_1^0$ is known, and thus also the macroscopic energy release rate $G = G(T)$, with $0 < G < k$. A blow-up of the solution around the tip of the process zone is implemented at the true cycle $i_d(T)$ through the introduction of the rescaled coordinate $y = (x - \lambda_{i_d(T)}^d)/d$. The following is shown to hold in (Jaubert and Marigo, 2006) for all $j \in \mathbb{Z}$ (in what follows, the dependence in T of G, and l_j defined below is implicit):

---

[16] adapted from the French Canada Dry Ginger Ale commercial

a. There exists $\dot{l}_j \in \mathbb{R}$ such that

$$\lim_{d \to 0} \frac{1}{d}(l^d_{i_d(T)+j} - l^d_{i_d(T)}) = \dot{l}_j; \qquad (9.13)$$

b. $\lim_{d \to 0} \frac{1}{d}(\lambda^d_{i_d(T)+j} - l^d_{i_d(T)+j}) = 2\eta\sqrt{\frac{G}{k}};$

c. For an arbitrary y, $\lim_{d \to 0} v^d_{i_d(T)+j}(\lambda^d_{i_d(T)} + d\,y)/d = \dot{v}(y - \dot{l}_j)$ with $\dot{v}$ defined by

$$\dot{v}(y) = \begin{cases} -\dfrac{G}{k} - \dfrac{y}{\eta}\sqrt{\dfrac{G}{k}} & \text{if} \quad y \le -2\eta\sqrt{\dfrac{G}{k}} \\[2ex] \dfrac{y^2}{4\eta^2} & \text{if} \quad -2\eta\sqrt{\dfrac{G}{k}} \le y \le 0 \\[2ex] 0 & \text{if} \quad y \ge 0 \end{cases}$$

and

d. The sequence $\{\dot{l}_j\}_{j \in \mathbb{Z}}$ satisfies the following family of non linear equations

$$\sum_{m=0}^{\infty}\left(2\eta\sqrt{\frac{G}{k}} - \dot{l}_j + \dot{l}_{j-m}\right)^{+2} = 4\eta^2. \qquad (9.14)$$

Then, the evolution is said to be *stationary* if there exists $\dot{l} > 0$, such that the sequence $\{\dot{l}_j = j\dot{l}\}_{j \in \mathbb{Z}}$ is a solution of (9.14). Note that for each potential energy release rate G, such that $0 < G < k$, *there exists a unique stationary regime*, given by the implicit non linear equation

$$\sum_{m=0}^{\infty}\left(2\eta\sqrt{\frac{G}{k}} - m\dot{l}\right)^{+2} = 4\eta^2. \qquad (9.15)$$

Given a stationary regime, $\dot{l}$ is given by (9.15) and conversely, a stationary regime can be associated with any solution $\dot{l}$ of (9.15) by $\dot{l}_j = j\dot{l}$, $j \in \mathbb{Z}$. But (9.15) admits a unique solution. Indeed, the function

$$(0, +\infty) \ni \dot{l} \mapsto \sum_{m=0}^{\infty}\left(2\eta\sqrt{\frac{G}{k}} - m\dot{l}\right)^{+2} - 4\eta^2$$

is strictly decreasing from $+\infty$ to $-4\eta^2(1 - G/k) < 0$ when $\dot{l}$ goes from 0 to $2\eta\sqrt{G/k}$, and it is constant and equal to $-4\eta^2(1 - G/k) < 0$ as soon as $\dot{l} \ge 2\eta\sqrt{G/k}$. Hence it only vanishes once.

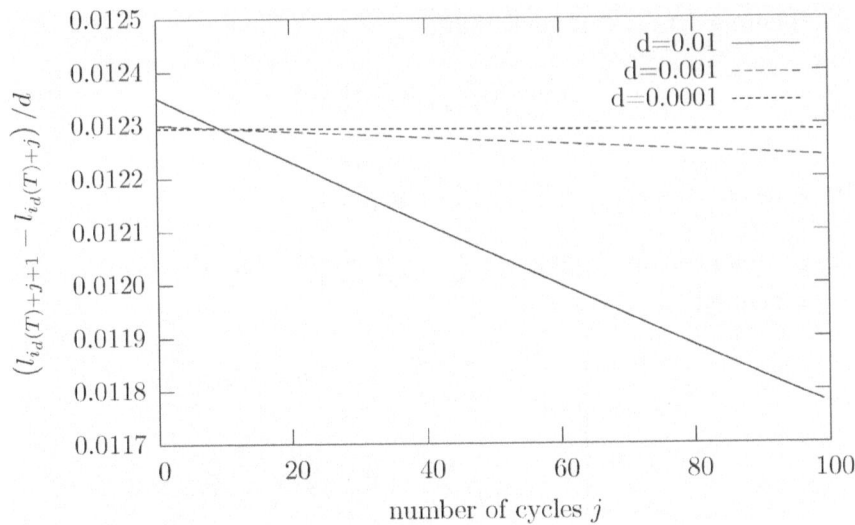

Figure 9.5. Numerical check of stationarity

REMARK 9.6. If the evolution is stationary, then in view of (9.13) – with $i_d(T) + j$ rewritten as $i_d(T + dj)$ – a diagonalization argument would show, as in Remark 9.5, that the quantity $\dot{l}(T)$ is also $\dfrac{dl}{dT}(T)$, provided the latter exists.

We will assume henceforth that

(Stat) *The stationary regime is the unique solution of* (9.14).

On Figure 9.2, for a set value $T = 15$, we plot $\big(l^d_{i_d(T)+j+1} - l^d_{i_d(T)+j}\big)\big)/d$ versus j for different values of d. The numerical values are obtained by solving the "true" non linear system (9.8)-(9.9) by a Newton-Raphson method. As seen on the represented curves, the growth rate is almost constant, *i.e.*, independent of j, when $d = 10^{-4}$. The regime seems indeed to be asymptotically stationary.

We now propose to establish a few simple properties of the limit law $\dot{l} = \mathsf{f}(G)$, the unique solution of (9.15). This is the object of the following

PROPOSITION 9.7.

1. $\dot{l}$ *cannot exist unless* $G \leq k$;

2. *When* $G = k$ *any value of* $\dot{l} \geq 2\eta$ *is solution of* (9.15);

3. When $0 < G < k$, f is continuously differentiable and increases from 0 to 2η when G goes from 0 to k;

4. Set $G_n := \dfrac{6nk}{(n+1)(2n+1)}$. In the interval $[G_{n+1}, G_n)$, $f(G)$ is given by

$$f(G) = \frac{6\eta}{2n+1}\sqrt{\frac{G}{k}} - \frac{2\eta}{2n+1}\sqrt{\frac{6(2n+1)}{n(n+1)} - \frac{3(n+2)}{n}\frac{G}{k}}. \quad (9.16)$$

Proof. The first item is immediate, once it is noted that equation (9.15) also has the form

$$\sum_{j=1}^{\infty}\left(2\eta\sqrt{\frac{G}{k}} - j\dot{i}\right)^{+2} = 4\eta^2\left(1 - \frac{G}{k}\right).$$

When $G = k$, the right hand side vanishes and the left hand side equals 0 if and only if $\dot{i} \geq 2\eta$, hence the second item.

When $0 < G < k$, we have previously established that (9.15) admits a unique solution $\dot{i} := f(G)$. For $G \in (0, k)$ and $\dot{i} > 0$, define F by

$$F(G, \dot{i}) := \sum_{j=0}^{\infty}\left(2\eta\sqrt{\frac{G}{k}} - j\dot{i}\right)^{+2} - 4\eta^2.$$

F is continuously differentiable, strictly increasing in G, given $\dot{i}$, and strictly decreasing in $\dot{i}$, given G. The implicit function theorem implies that f is continuously differentiable and increasing. For a given G, define $n(G)$ so that

$$\sum_{j=0}^{n(G)}\left(2\eta\sqrt{\frac{G}{k}} - j\dot{i}\right)^{+2} = 4\eta^2.$$

Then, $n(G) = n$ when

$$\frac{2\eta}{n+1}\sqrt{\frac{G}{k}} \leq f(G) < \frac{2\eta}{n}\sqrt{\frac{G}{k}}. \quad (9.17)$$

Consequently,

$$4\frac{G}{G_n} - 4 = F\left(G, \frac{2\eta}{n}\sqrt{\frac{G}{k}}\right) < 0 = F(G, f(G))$$

$$\leq F\left(G, \frac{2\eta}{n+1}\sqrt{\frac{G}{k}}\right) = 4\frac{G}{G_{n+1}} - 4.$$

In other words, to find $n(G)$, it is enough to determine the interval $[G_{n+1}, G_n)$ in which G lies. Once $n(G)$ is found, $\dot{l}$ is given as root of the following quadratic equation:

$$
\begin{aligned}
0 &= \sum_{j=0}^{n(G)} \left(2\eta\sqrt{\frac{G}{k}} - j\dot{l} \right)^2 - 4\eta^2 \\
&= \frac{1}{6}n(G)(n(G)+1)(2n(G)+1)\dot{l}^2 - 2\eta n(G)(n(G)+1)\sqrt{\frac{G}{k}}\,\dot{l} \\
&\quad + 4\eta^2(n(G)+1)\frac{G}{k} - 4\eta^2.
\end{aligned}
$$

The only relevant root is such that (9.17) is satisfied, hence (9.16). $\quad\square$

The graph of the function f is plotted on Figure 9.6.

When $G \nearrow k = G_1$, then $n(G) = 1$ and $\dot{l} \to 2\eta$. If G is near k, $n(G) = 1$ and

$$
\mathsf{f}(G) = 2\eta\sqrt{\frac{G}{k}} - 2\eta\sqrt{1 - \frac{G}{k}}.
$$

The slope of f is infinite at $G = k$, the graph of f becomes tangent to the half-line $[2\eta, \infty)$, i.e., the set of solutions of (9.15) when $G = k$.

When $G \searrow 0$, then $n(G) \to \infty$ and $\dot{l} \to 0$. Actually, the behavior near 0 is described through the following

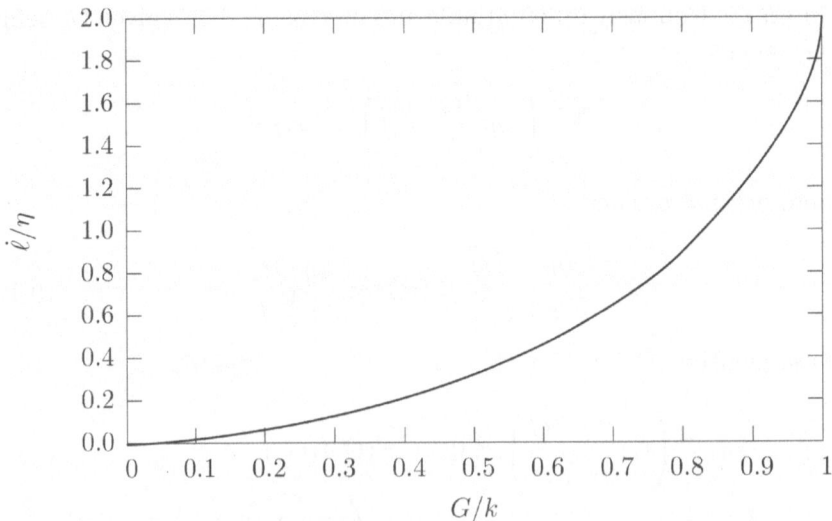

Figure 9.6. Graph of the fatigue limit law $\mathsf{f}(G)$

REMARK 9.8. When G/k is small, the fatigue limit law is like a Paris law with exponent $3/2$, that is

$$f(G) \approx \frac{2\eta}{3} \left(\frac{G}{k} \right)^{\frac{3}{2}}. \qquad (9.18)$$

Indeed, from $G_n/k = 6n/(n+1)(2n+1)$, we get $n(G) \approx 3k/G$. Inserting this into (9.16) yields (9.18).

We recall (Stat) from page 144, Remarks 9.5, 9.6, together with the fact that $G(T)$ is given by Proposition 9.4. This establishes the following

PROPOSITION 9.9. *Under the a priori assumption that $l(T)$ is differentiable, the evolution $T \in (0, +\infty) \mapsto l(T)$ of the debond length satisfies*

$$\frac{dl}{dT}(T) = f(G(T))$$

$$G(T) = \frac{\eta^2 \overline{V}^2}{l(T)^2} k$$

$$l(0+) = \eta \overline{V}.$$

The mechanically versed reader cannot fail to recognize in the above result a typical fatigue law à la Paris, and in (9.18) a typical functional shape for such a law, albeit one "genitum non factum"[17], in contrast to what is, to the best of our knowledge, currently available in the existing literature.

The next subsection attempts to incorporate the result obtained in Proposition 9.9 in the general framework developed in Section 2, and, specifically, in Subsection 2.4.

9.3. A VARIATIONAL FORMULATION FOR FATIGUE

9.3.1. *Peeling revisited*

First, we shall view T as the time variable and $\overline{V}$ as a load applied as soon as $T = 0+$. This instantaneous load (with respect to the variable T) generates a jump (debonding) at $x = 0$, that given by Griffith's model at the end of the loading part of the first cycle, *i.e.*, according to Proposition 9.9, $\eta \overline{V}$.

In view of the first item in Proposition 9.7, the fatigue law obtained in Proposition 9.9 encompasses both a Paris type law and a Griffith

---

[17] Council of Nicea −325 A.D.

148   B. Bourdin, G. Francfort and J.-J. Marigo

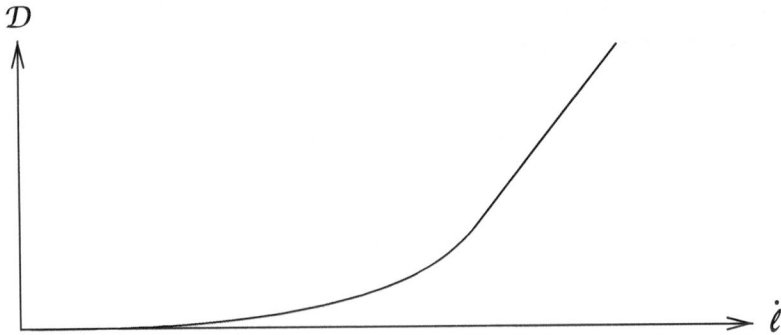

Figure 9.7. Dissipation potential $\mathcal{D}$

type law. We now rewrite it, in the thermodynamical language of e.g.
(Halphen and Nguyen, 1975), as a generalized standard law of the form

$$\dot{l} \in \partial \mathcal{D}^*(G) \qquad \text{or} \qquad G \in \partial \mathcal{D}(\dot{l}),$$

where $\mathcal{D}$ is convex and $\mathcal{D}^*$ its conjugate. Specifically,

$$\mathcal{D}(\dot{l}) = \begin{cases} \displaystyle\int_0^{\dot{l}} \mathsf{f}^{-1}(\lambda)d\lambda & \text{if } \dot{l} \leq 2\eta \\ \displaystyle\int_0^{2\eta} \mathsf{f}^{-1}(\lambda)d\lambda + k(\dot{l} - 2\eta) & \text{if } \dot{l} \geq 2\eta. \end{cases} \tag{9.19}$$

The dissipation potential $\mathcal{D}$ is linear with slope k at large enough
propagation speeds, and, because of (9.18), follows a power law with
exponent 5/3 for small propagation speeds (see Figure 9.7).

An equivalent statement of the generalized standard law is that $\dot{l}$
is a minimizer, among non-negative λ's, for $\mathcal{D}(\lambda) - G\lambda$. By virtue of
(9.12), discretizing time – with time increments ΔT – then permits us
to write the following discretized version of the generalized standard
law:

$$\frac{\Delta l}{\Delta T} = \operatorname{argmin}_{\lambda \geq 0} \left\{ \frac{1}{\Delta T}(\mathcal{P}(l + \lambda\Delta T) - \mathcal{P}(l)) + \mathcal{D}(\lambda) \right\},$$

or still, denoting by l_I the debond length at the Ith time step, and
setting $l_0 = 0$,

$$l_{I+1} = \operatorname{argmin}_{\{l \geq l_I\}} \left\{ \mathcal{P}(l) + \Delta T \, \mathcal{D}\left(\frac{l - l_I}{\Delta T}\right) \right\}. \tag{9.20}$$

The latter formulation falls squarely within the setting developed in
Subsection 2.4. Fatigue is indeed an evolution problem of the type

discussed throughout, albeit for a non 1-homogeneous dissipation potential.

9.3.2. *Generalization*

Peeling is but one example for which the above analysis can be carried out. Of course, more complex settings may map into hostile terrain which would jeopardize the analytical subtleties that were key to the successful completion of the analysis of the second order limit in the peeling case. The ensuing dissipation potential for the limit fatigue problem will then be completely out of reach. However, we do expect similar qualitative behavior, at least for preset crack paths and simple cyclic loads.

In the absence of a well-defined crack path, the hostility scale tips towards chaos. First, the parameter l must be replaced by the crack set Γ at a given time. The add-crack at each time step is of the form $\Gamma\backslash\Gamma_I$.

The potential energy $\mathcal{P}(\Gamma)$ associated with Γ is still obtained as an elasticity problem on the cracked structure submitted to the maximal amplitude load (see (9.12)), but the computation of the dissipation potential becomes very tricky. In particular, each connected component of the add-crack must be counted separately in that computation, because, since that potential is no longer linear in the length, a different result would be achieved when considering e.g. a structure made of two identical connected components, both submitted to the same load, according to whether we view the the resulting two cracks as one crack or two cracks!

Consequently, the dissipation potential becomes

$$\Delta T \sum_{k\in K(\Gamma\backslash\Gamma_I)} \mathcal{D}\Big(\frac{\mathcal{H}^1((\Gamma\backslash\Gamma_I)^k)}{\Delta T}\Big),$$

where $K(\gamma)$ denotes the number of connected components of a set γ and γ^k its kth-component. The generalized incremental problem may thus be stated as

Find Γ_{I+1} (local) minimizer on $\{\Gamma : \Gamma \supset \Gamma_I\}$ of

$$\mathcal{P}(\Gamma) + \Delta T \sum_{k\in\mathcal{K}(\Gamma\backslash\Gamma_I)} \mathcal{D}\Big(\frac{\mathcal{H}^1((\Gamma\backslash\Gamma_I)^k)}{\Delta T}\Big)$$

with Γ_0 given.

Summary dismissal of the above formulation on the grounds of vagueness cannot be argued against at present. The authors readily concede that Paragraph 9.3.2 amounts to little more than a discourse on "known unknowns"[18]; but, in all fairness, this alone is a marked improvement over the "unknown unknowns"[19] of classical fatigue.

---

[18] Donald Rumsfeld – Feb. 12, 2002, U.S. Department of Defense news briefing
[19] idem

Appendix

As announced in the Introduction, this short Appendix consists of a self-contained, but succinct exposition of the necessary mathematical prerequisites for a successful reading of the material presented in this tract. Redundancy, while not favored, is not avoided either in that quite a few definitions or properties may be found here as well as in the main body of the text. We thus favor fluidity over strict logics.

Two measures on $\mathbb{R}^N$ play a pivotal role throughout the text, the Lebesgue measure, denoted by dx, and the $(N-1)$-Hausdorff measure, denoted by $\mathcal{H}^{N-1}$, and defined in e.g. (Evans and Gariepy, 1992), Section 2.1. The unfamiliar reader may think of the latter as coinciding with the $(N-1)$-surface measure on smooth enough hypersurfaces.

As in most papers in applied analysis, derivatives are generically weak derivatives; these are meant as distributional derivatives, which make sense as soon as the field v that needs differentiation is locally integrable. The weak derivative is denoted by ∇v. Also, we systematically use weak and/or weak-* convergence (both being denoted by $\rightharpoonup$), appealing to the following weak version of Banach-Alaoglu's theorem found in e.g. (Rudin, 1973), Theorem 3.17:

THEOREM A. *If X is a separable Banach space, then any bounded sequence in X^* has a weak-* converging subsequence.*

Spaces. Denoting by $B(x, \rho)$ the ball of center x and radius ρ, we recall that a Lebesgue point of a function $u \in L^1_{loc}(\mathbb{R}^N)$ is a point x such that

$$\lim_{\rho \downarrow 0} \frac{1}{|B(x,\rho)|} \int_{B(x,\rho)} |u(y) - u(x)| \, dx = 0,$$

and that Lebesgue-almost every point of $\mathbb{R}^N$ is a Lebesgue point for u.

The space X will often be a Sobolev space of the form $W^{1,p}(\Omega)$, with Ω a bounded (connected) open set of $\mathbb{R}^N$ and $1 < p \leq \infty$. We adopt the following definition for $W^{1,p}(\Omega)$:

$$W^{1,p}(\Omega) = \left\{ v \in L^p(\Omega) : \nabla v \in L^p(\Omega; \mathbb{R}^N) \right\},$$

and use the notation $W^{1,p}(\Omega; \mathbb{R}^m)$ for $\mathbb{R}^m$-valued functions whose components are in $W^{1,p}(\Omega)$. The same applies to $(S)BV(\Omega; \mathbb{R}^m)$; see below.

We will use the classical imbedding and compactness theorems for Sobolev spaces – see e.g. (Adams and Fournier, 2003), Theorems 4.12 and 6.3 – most notably that, provided that the boundary of Ω is smooth enough, say Lipschitz, then, for $p < N$,

$$W^{1,p}(\Omega) \to L^q(\Omega), \text{ resp. } W^{1,p}(\Omega) \hookrightarrow L^q(\Omega); \ 1/q \geq \text{ resp. } > 1/p - 1/N,$$

where $\hookrightarrow$ stands for "compact embedding".

The space $BV(\Omega)$ is of particular relevance to the study of fracture. It is defined as

$$BV(\Omega) = \left\{ v \in L^1(\Omega) : \sup_\varphi \left\{ \int_\Omega u \operatorname{div} \varphi \, dx : \varphi \in \mathcal{C}_0^\infty(\Omega; \overline{B}(0,1)) \right\} < \infty \right\}.$$

Thanks to the Riesz representation theorem (see e.g. (Evans and Gariepy, 1992), Section 1.8), for any $u \in BV(\Omega)$, there exists a non-negative bounded Radon measure μ and a μ-measurable function σ with $\sigma(x) = 1$, μ-a.e. such that $\nabla u = \sigma \mu$. The variation measure μ is denoted by $|Du|$ and $|Du|(\Omega)$ is the total variation of u.

The beauty of BV-functions is epitomized by the following structure theorem (see e.g. (Ambrosio et al., 2000), Sections 3.7-3.9):

THEOREM B. *Consider $u \in BV(\Omega)$. Then*

$$Du = \nabla u \, dx + (u^+ - u^-)\nu \, \mathcal{H}^1 \lfloor S(u) + C(u),$$

where

- $\nabla u \in L^1(\Omega; \mathbb{R}^N)$ *is the approximate differential of u, i.e.,*

$$\lim_{\rho \downarrow 0} \frac{1}{|B(x,\rho)|} \int_{B(x,\rho)} \frac{|u(y) - u(x) - \nabla u(x).(y-x)|}{\rho} dy = 0;$$

- $S(u)$ *is the complement of the set of Lebesgue points for u, a countably 1-rectifiable set, i.e., the countable union of compact subsets of $\mathcal{C}^1$-hypersurfaces, up to a set of $\mathcal{H}^{N-1}$-measure 0;*

- $\nu(x)$ *is the normal at a point x of $S(u)$ to that set;*

- $\displaystyle \lim_{\rho \downarrow 0} \frac{1}{|B_\nu^\pm(x,\rho)|} \int_{B_\nu^\pm(x,\rho)} |u(y) - u^\pm(x)| \, dy = 0$, *with*

$$B_\nu^\pm(x,\rho) := \{ y : (y-x).\nu \in (0, \pm\rho) \};$$

and

- $|C(u)|(B) = 0$ *if $\mathcal{H}^{N-1}(B) < \infty$; $C(u)$ and dx are mutually singular.*

Finally $|Du(\Omega)| = \int_\Omega |\nabla u| dx + \int_{S(u)} |u^+ - u^-| d\mathcal{H}^{N-1} + |C(u)|(\Omega)|.$

$BV(\Omega)$ enjoys the following injection and compactness properties, for Lipschitz bounded Ω's:

$$BV(\Omega) \to L^{\frac{N}{N-1}}(\Omega); \quad BV(\Omega) \hookrightarrow L^p(\Omega), \ p < \frac{N}{N-1}.$$

The Cantor part $C(u)$, which can be seen as a diffuse measure should be *a priori* avoided in fracture, so that the correct space for fracture is

$$SBV(\Omega) := \{v \in BV(\Omega) : C(v) = 0\}.$$

That space is manageable because of Ambrosio's compactness theorem, detailed on page 24, which states a compactness result for weak-* convergence in $SBV(\Omega)$.

Minimization. The so-called direct method of the Calculus of Variations always revolves around some variant of the same basic minimization result, namely,

THEOREM C. *Consider X a reflexive Banach space, or the dual of a separable Banach space. Let $I : X \mapsto \overline{\mathbb{R}}$ be such that*

- *I is weak(-*) lower semi-continuous; and*

- $\limsup_n I(u_n) = \infty$ *when* $\|u_n\| \overset{n}{\nearrow} \infty$.

Then, I admits a minimizer over X.

Recall that, if I is convex, then, for any minimizer u^* of I, 0 is in the sub-differential of I at u^*,

$$0 \in \partial I(u^*),$$

where, for any $v \in X$,

$$\partial I(v) := \{v^* \in X^* : I(w) - I(v) \geq \langle v^*, w - v \rangle, \forall w \in X\}.$$

For functionals of the form $I(u) := \int_\Omega W(\nabla u)\, dx$ with $\Omega \in \mathbb{R}^N$ and $W : \mathbb{R}^N \mapsto \mathbb{R}$ nonnegative and continuous, convexity is equivalent to to lower semi-continuity in the scalar case, *i.e.*, whenever $u : \Omega \mapsto \mathbb{R}$ or when $N = 1$, as explained in e.g. (Dacorogna, 1989).

If departing from the scalar case, one should replace convexity of W with a less pleasant notion, that of quasiconvexity; see (Dacorogna, 1989). A continuous non-negative functional $W : \mathbb{R}^{N^2} \mapsto \mathbb{R}$ is quasi-convex iff, for any $F \in \mathbb{R}^{N^2}$,

$$W(F) \leq \inf_\varphi \left\{ \int_C W(F + \nabla\varphi)\, dx : \varphi \in W_0^{1,\infty}(\Omega; \mathbb{R}^N) \right\},$$

where C stands for the unit cube centered at 0. Note that the definition is independent of the choice of the base domain C – which can be replaced by any bounded open set – and also note that, for the choice of C as base domain, $W_0^{1,\infty}(\Omega; \mathbb{R}^N)$ can be replaced by $W_{per}^{1,\infty}(\Omega; \mathbb{R}^N)$, as seen in (Ball and Murat, 1984).

So, except in the anti-plane shear case, quasi-convexity of the bulk energy will replace convexity for functionals of $\nabla\varphi$, with $\varphi : \Omega \mapsto \mathbb{R}^{2,3}$, the deformation field.

In this context, various results of Ambrosio, found in e.g. (Ambrosio et al., 2000), Section 5.4, lead to the following

THEOREM D. *Let* $\phi(i,j,p) := \gamma(|i-j|)\psi(p)$, *with*

- $i,j \in K$, *compact of* $\mathbb{R}^N$, $p \in \mathbb{R}^N$;

- γ *lower semi-continuous, increasing and sub-additive* (*i.e.,* $\gamma(i + j) \le \gamma(i) + \gamma(j)$);

- ψ *even, convex and positively 1-homogeneous.*

Let $W : \mathbb{R}^{N^2} \mapsto \mathbb{R}$ *be continuous and satisfy*

$$1/\mathcal{C}|F|^p \le W(F) \le \mathcal{C}(1 + |F|^p), \ 1 < p < \infty.$$

Then, the functional

$$I(u) := \int_\Omega W(\nabla u)dx + \int_{S(u)} \phi(u^+, u^-, \nu)d\mathcal{H}^{N-1}, \ u \in SBV(\Omega; \mathbb{R}^N),$$

is lower semi-continuous on $SBV(\Omega; \mathbb{R}^N)$,*i.e.,*

$$I(u) \le \liminf_n I(u_n)$$

whenever $u_n \in SBV(\Omega; \mathbb{R}^N)$ *converges strongly in* $L^1(\Omega; \mathbb{R}^N)$ *to* $u \in SBV(\Omega; \mathbb{R}^N)$, *with* $\mathcal{H}^{N-1}(S(u_n) \le \mathcal{C}' < \infty$.

This theorem leads to the existence of a minimizer for the weak discrete evolution (Wde) defined in Paragraph 5.1.1. It suffices to note that $\gamma(t) = \begin{cases} 0, \ t = 0 \\ 1, \ \text{otherwise} \end{cases}$ and $\psi \equiv 1$ satisfy the assumptions of Theorem D.

Now, whenever a functional I is not lower semi-continuous, then, in the context of Theorem C,

$$\inf I(u) = \min I^*(u),$$

where I^* is the lower semi-continuous envelope of I, *i.e.,* the greatest lower semi-continuous functional below I. It is defined as

$$I^*(u) := \inf\left\{\liminf_n I(u_n) : \{u_n\} \text{ such that } u_n \rightharpoonup u\right\}.$$

In the context of functionals of the form $I(u) := \int_\Omega W(\nabla u) \, dx$ with $W : \mathbb{R}^{N^2} \mapsto \mathbb{R}$ nonnegative, continuous and such that $W(F) \leq C(1+|F|^p)$, $1 \leq p < \infty$, it was established in (Acerbi and Fusco, 1984), Statement [III.7], that

$$I^*(u) = \int_\Omega QW(\nabla u) dx,$$

where QW, the quasiconvexification of W is defined as

$$QW(F) := \inf_\varphi \left\{ \int_C W(F + \nabla\varphi) \, dx : \varphi \in W_0^{1,\infty}(\Omega; \mathbb{R}^N) \right\}.$$

For functionals defined on $BV(\Omega; \mathbb{R}^N)$, the results are more recent and are evoked on page 68.

Approximation. The computation of minimizers for a functional of the Mumford-Shah type is not trivial, as illustrated in Section 8. The idea is to approximate the functional by a sequence of functionals, such that the corresponding minimizers converge to a minimizer of the original functional. The relevant definition is that of Γ-convergence given on page 108.

The following equivalent definition of Γ-convergence is very useful (see (Braides, 2002), Remark 1.6):

THEOREM E. *In the notation of page 108, $\mathcal{F}_\varepsilon$ Γ-converges to $\mathcal{F}$ iff*

$$\Gamma - \liminf \mathcal{F}_\varepsilon = \Gamma - \limsup \mathcal{F}_\varepsilon,$$

with

$\Gamma - \liminf \mathcal{F}_\varepsilon(u)$ (*resp.* $\Gamma - \limsup \mathcal{F}_\varepsilon(u)) :=$

$\inf \{ \liminf_\varepsilon \mathcal{F}_\varepsilon(u_\varepsilon)$ (*resp.* $\limsup_\varepsilon \mathcal{F}_\varepsilon(u_\varepsilon)) : \{u_\varepsilon\}$ *such that* $u_\varepsilon \to u \}.$

From the standpoint of Γ-convergence, computing the lower semi-continuous envelope of a functional I is the same as computing the $\Gamma - \liminf$ of $\mathcal{F}_\varepsilon$ with $\mathcal{F}_\varepsilon \equiv \mathcal{F}$!

We do not dwell any further on that notion, noting however that, in a separable metric space, any sequence $\mathcal{F}^\varepsilon$ of $\overline{\mathbb{R}}$-valued functionals admits a Γ-converging subsequence.

Glossary

$BV(\Omega)$ the space of functions defined on Ω with bounded variation, Appendix, 24, 152

F the deformation gradient, 9, 18, 28, 58, 73, 80, 104–106, 110, 153, 154

G the energy release rate. In 2d, it is minus the derivative of the potential energy of the body with respect to the crack length, 1, 5, 12, 47, 56, 90, 123, 141–143

$G \in \partial\mathcal{D}(\dot{l})$ reads G *belongs to the subdifferential of* $\mathcal{D}$ *at* $\dot{l}$. The notion of sub-differential generalizes that of derivative for non differentiable but convex function. Formally, it is equivalent to the inequality $G(\lambda - \dot{l}) \leq \mathcal{D}(\lambda) - \mathcal{D}(\dot{l})$, $\forall\lambda$, Appendix, 16, 153

$L^p(\Omega)$ the space of functions defined on Ω and whose p-th power is Lebesgue-integrable, 73, 151

$S(\varphi)$ the jump set of φ, *i.e.*, the set of points where φ is discontinuous, Appendix, 24

$SBV(\Omega)$ the subspace of functions of $BV(\Omega)$ with no Cantor part, Appendix, 24, 152

$W(F)$ the bulk (elastic) energy density, 9, 18, 28, 58, 73, 80, 104, 106, 110, 154

$W^{1,p}(\Omega)$ the subspace of functions of $L^p(\Omega)$ the first weak derivative of which is also in $L^p(\Omega)$, Appendix, 73, 151

$[\varphi]$ the jump of φ, Appendix, 27

$\Gamma(l)$ the crack set corresponding to a crack of length l in the case of a predefined crack path, 11–13, 15, 17, 19, 21, 50, 51, 56

$\Gamma(t)$ the crack state at time t, *i.e.*, roughly the set of material points where the deformation φ is or has been discontinuous, 22

$\hat{\Gamma}$ the predefined crack path, *i.e.*, the given set of material points where the deformation φ can be discontinuous, 11, 22

$\hookrightarrow$ compact embedding, Appendix, 151

κ the surface energy density for cohesive force models. In the case of the Dugdale model: $\kappa(\delta) = \max\{\sigma_c\delta, k\}$, 27

Eb the energy balance. It is a principle of conservation of the total energy of the body. This requirement complements the unilateral condition (Ust), (Ulm) or (Ugm). Their concatenation induces the propagation law for the evolution of the crack, 15, 17, 25, 28, 88

Griffith's theory the theory of fracture mechanics in which the surface energy is proportional to the area of the the crack and the propagation law is based on the critical energy release rate criterion, 4, 10

Sde the strong discrete evolution. After time discretization, the minimization problem which yields the crack Γ_i and the deformation φ_i at each step i, 70, 74, 75, 90

Strong formulation the variational formulation is said to be strong when the deformation and the crack are considered as independent variables, 23, 26, 51, 52, 54, 55, 70–75, 81, 82, 90

Ugm Unilateral global minimality condition : it requires that the given state be that with the smallest energy among all admissible states with a larger crack state. The corresponding mathematical statement depends on whether the formulation is weak or strong, on whether the crack path is prescribed or free, and on whether the surface energy is of a Griffith or cohesive kind, 19, 22, 25, 28, 88

Ulm Unilateral local minimality condition : it requires that the given state have the smallest total energy among all admissible states in its neighborhood with a larger crack set. The corresponding mathematical statement depends on whether the formulation is weak or strong, on whether the crack path is prescribed or free, and on whether the surface energy is of a Griffith or cohesive kind. It is also topology dependent in that the neighborhood refers to a specific topology, 19, 22, 25, 28, 87

Ust Unilateral stationarity condition: it requires that the first derivative of the total energy at an actual state be non negative in any admissible direction that increases the crack set in the body. The corresponding mathematical statement depends on whether the formulation is weak or strong, on whether the crack path is prescribed or free, and on whether the surface energy is of a Griffith or cohesive kind, 15–17

Wde the weak discrete evolution. After time discretization, the minimization problem which yields the deformation φ_i at each step i, 70, 74, 75, 77, 78, 90, 109, 154

Weak formulation the variational formulation is said to be weak when the configuration is the unique independent variable, the crack being considered as the set where the deformation is or has been discontinuous, viii, 3, 17, 22, 25, 26, 28, 31, 33, 34, 37, 40, 48, 49, 51, 55, 67, 70, 72, 74, 75, 77, 80, 82, 87, 88, 90, 93, 104–106, 109, 114, 154

the cardinal of a set, 8, 32, 33, 35

cumulated opening the variable memorizing the additional opening of the crack at a given point, up to the present time. It is used to enforce irreversibility in a cohesive model, 28, 85, 136

debonded zone the part of the crack lips where the cohesive forces vanish. In the Dugdale model, it corresponds to the points where the (cumulated) opening is greater than d, 137–139, 141

global minimizer the global or absolute minimizer of a real-valued function f over a set X is the smallest value that the function takes on the whole set. This concept is independent of the ambient topology in the set X, viii, 1, 3–7, 19–23, 25, 26, 28, 31–33, 42, 43, 48, 49, 51, 52, 56, 58, 67, 70, 74, 78, 83, 88, 90, 103, 107, 108, 114, 117, 118, 123, 125, 130

local minimum a real-valued function f admits a local or relative minimum at a point x of a set X, if there exists a neighborhood of x in which $f(x)$ is a minimum. This concept depends on the ambient topology of X, viii, 3–6, 19, 20, 22, 23, 25, 28, 31–33, 36, 39, 41, 42, 51, 55–58, 70, 87, 103, 113, 115, 117–119, 129, 149

maximal opening the variable memorizing the maximal value of the crack opening at a given point, up to the present time. It is used to enforce irreversibility in a cohesive model, 83, 85, 86

opening the jump of the normal displacement across the crack, 85, 86, 136, 138, 140, 141

process zone or **cohesive zone** the part of the crack lips where the cohesive forces are active. In the Dugdale model, it corresponds to the points where the (cumulated) opening is less than d, 5, 47, 106, 137–139, 141

unilateral In the definitions (Ust), (Ulm) and (Ugm), the word unilateral is introduced so as to emphasize that the tested configuration has to be compared only to configurations with a larger crack set. Moreover, the stationarity condition in (Ust) is an inequality and can be seen as a first order optimality condition, 3, 15, 20

References

Abdelmoula, R. and J.-J. Marigo: 2000, 'The effective behavior of a fiber bridged crack'. *J. Mech. Phys. Solids* **48**(11), 2419–2444.

Acerbi, E. and N. Fusco: 1984, 'Semicontinuity Problems in the Calculus of Variations'. *Arch. Rat. Mech. Anal.* **86**(2), 125–145.

Adams, A. and J. Fournier: 2003, *Sobolev Spaces*, Pure and Applied Mathematics. Elsevier, 2nd edition.

Alberti, G.: 2000, 'Variational Models for Phase Transitions, an Approach via Γ–Convergence'. In: G. Buttazzo (ed.): *Calculus of Variations and Partial Differential Equations*. pp. 95–114, Springer–Verlag.

Ambrosio, L.: 1990, 'Existence theory for a new class of variational problems'. *Arch. Ration. Mech. An.* **111**, 291–322.

Ambrosio, L.: 1994, 'On the lower semicontinuity of quasiconvex integrals in $\text{SBV}(\Omega, \mathbf{R}^k)$'. *Nonlinear Anal.-Theor.* **23**(3), 405–425.

Ambrosio, L., N. Fusco, and D. Pallara: 2000, *Functions of Bounded Variation and Free Discontinuity Problems*. Oxford University Press.

Ambrosio, L. and V. Tortorelli: 1990, 'Approximation of functionals depending on jumps by elliptic functionals via Γ-convergence'. *Commun. Pur. Appl. Math.* **43**(8), 999–1036.

Ambrosio, L. and V. Tortorelli: 1992, 'On the approximation of free discontinuity problems'. *Boll. Un. Mat. Ital. B (7)* **6**(1), 105–123.

Ball, J. M. and F. Murat: 1984, '$W^{1,p}$-quasiconvexity and variational problems for multiple integrals'. *J. Funct. Anal.* **58**(3), 225–253.

Barenblatt, G. I.: 1962, 'The mathematical theory of equilibrium cracks in brittle fracture'. *Adv. Appl. Mech.* pp. 55–129.

Bellettini, G. and A. Coscia: 1994, 'Discrete approximation of a free discontinuity problem'. *Numer. Func. Anal. opt.* **15**(3-4), 201–224.

Bilteryst, F. and J.-J. Marigo: 2003, 'An energy based analysis of the pull-out problem'. *Eur. J. Mech. A-Solid* **22**, 55–69.

Bonnet, A. and G. David: 2001, 'Cracktip is a global Mumford-Shah minimizer'. *Astérisque* (274), vi+259.

Bouchitté, G., A. Braides, and G. Buttazzo: 1995, 'Relaxation results for some free discontinuity problems'. *J. Reine Angew. Math.* **458**, 1–18.

Bouchitté, G., I. Fonseca, G. Leoni, and L. Mascarenhas: 2002, 'A global method for relaxation in $W^{1,p}$ and in SBV^p'. *Arch. Ration. Mech. An.* **165**(3), 187–242.

Bourdin, B.: 1998, 'Une méthode variationnelle en mécanique de la rupture. Théorie et applications numériques'. Thèse de doctorat, Université Paris-Nord.

Bourdin, B.: 1999, 'Image segmentation with a finite element method'. *ESAIM Math. Model. Numer. Anal.* **33**(2), 229–244.

Bourdin, B.: 2006, 'The variational formulation of brittle fracture: numerical implementation and extensions'. In: A. Combescure, T. Belytschko, and R. de Borst (eds.): *IUTAM Symposium on Discretization Methods for Evolving Discontinuities*. pp. 381–394. Springer.

Bourdin, B.: 2007a, 'Numerical implementation of a variational formulation of quasi-static brittle fracture'. *Interfaces Free Bound.* **9**, 411–430.

Bourdin, B. and A. Chambolle: 2000, 'Implementation of an adaptive finite-element approximation of the Mumford-Shah functional'. *Numer. Math.* **85**(4), 609–646.

Bourdin, B., G. A. Francfort, and J.-J. Marigo: 2000, 'Numerical experiments in revisited brittle fracture'. *J. Mech. Phys. Solids* **48**, 797–826.

Braides, A.: 2002, Γ-*convergence for Beginners*, Vol. 22 of *Oxford Lecture Series in Mathematics and its Applications*. Oxford University Press.

Braides, A., G. Dal Maso, and A. Garroni: 1999, 'Variational formulation of softening phenomena in fracture mechanics: the one dimensional case.'. *Arch. Ration. Mech. An.* **146**, 23–58.

Bucur, D. and N. Varchon: 2000, 'Boundary variation for a Neumann problem'. *Ann. Scuola Norm.-Sci.* **29**(4), 807–821.

Bui, H. D.: 1978, *Mécanique de la Rupture Fragile*. Paris: Masson.

Chaboche, J.-L., F. Feyel, and Y. Monerie: 2001, 'Interface debonding models: a viscous regularization with a limited rate dependency'. *Int. J. Solids Struct.* **38**(18), 3127–3160.

Chambolle, A.: 2003, 'A density result in two-dimensional linearized elasticity, and applications'. *Arch. Ration. Mech. An.* **167**(3), 211–233.

Chambolle, A.: 2004, 'An approximation result for special functions with bounded variations'. *J. Math. Pure Appl.* **83**, 929–954.

Chambolle, A.: 2005, 'Addendum to "An approximation result for special functions with bounded deformation" [J. Math. Pures Appl. (9) 83 (7) (2004) 929–954]: the N-dimensional case'. *J. Math. Pure Appl.* **84**, 137–145.

Chambolle, A. and F. Doveri: 1997, 'Continuity of Neumann linear elliptic problems on varying two-dimensional bounded open sets'. *Commun. Part. Diff. Eq.* **22**(5-6), 811–840.

Chambolle, A., A. Giacomini, and M. Ponsiglione: 2007, 'Crack initiation in brittle materials'. *Arch. Ration. Mech. An.* DOI: 10.1007/s00205-007-0080-6.

Charlotte, M., G. A. Francfort, J.-J. Marigo, and L. Truskinovky: 2000, 'Revisiting brittle fracture as an energy minimization problem: comparison of Griffith and Barenblatt surface energy models'. In: A. Benallal (ed.): *Continuous Damage and Fracture*. Paris, pp. 7–12, Elsevier.

Charlotte, M., J. Laverne, and J.-J. Marigo: 2006, 'Initiation of cracks with cohesive force models: a variational approach'. *Eur. J. Mech. A-Solid* **25**(4), 649–669.

Ciarlet, P. G.: 1986, *Élasticité tridimensionnelle*, Vol. 1 of *Recherches en Mathématiques Appliquées [Research in Applied Mathematics]*. Paris: Masson.

Ciarlet, P. G. and J. Nečas: 1987, 'Injectivity and self-contact in nonlinear elasticity'. *Arch. Ration. Mech. An.* **97**(3), 171–188.

Dacorogna, B.: 1989, *Direct Methods in the Calculus of Variations*. Berlin, Heidelberg: Springer Verlag.

Dal Maso, G.: 1993, *An introduction to Γ-convergence*. Boston: Birkhäuser.

Dal Maso, G., A. De Simone, M. G. Mora, and M. Morini: 2006, 'A vanishing viscosity approach to quasistatic evolution in plasticity with softening'. Technical report, SISSA.

Dal Maso, G., G. A. Francfort, and R. Toader: 2005, 'Quasistatic crack growth in nonlinear elasticity'. *Arch. Ration. Mech. An.* **176**(2), 165–225.

Dal Maso, G. and R. Toader: 2002, 'A Model for the Quasi-Static Growth of Brittle Fractures: Existence and Approximation Results'. *Arch. Ration. Mech. An.* **162**, 101–135.

Dal Maso, G. and C. Zanini: 2007, 'Quasistatic crack growth for a cohesive zone model with prescribed crack path'. *Proc. Roy. Soc. Edin. A* **137**(2), 253–279.

De Giorgi, E., M. Carriero, and A. Leaci: 1989, 'Existence theorem for a minimum problem with free discontinuity set'. *Arch. Ration. Mech. An.* **108**, 195–218.

Del Piero, G.: 1997, 'One dimensional ductile-brittle transition, yielding, and structured deformations'. In: P. Argoul and M. Frémond (eds.): *Proceedings of the*

IUTAM Symposium "Variations de domaines et frontières libres en mécanique". pp. 197–202, Kluwer.

Destuynder, P. and M. Djaoua: 1981, 'Sur une interprétation mathématique de l'intégrale de Rice en théorie de la rupture fragile'. *Math. Met. Appl. Sc.* **3**, 70–87.

Dugdale, D. S.: 1960, 'Yielding of steel sheets containing slits'. *J. Mech. Phys. Solids* **8**, 100–108.

Evans, L. and R. Gariepy: 1992, *Measure theory and fine properties of functions*. Boca Raton, FL: CRC Press.

Federer, H.: 1969, *Geometric measure theory*. New York: Springer-Verlag.

Ferriero, A.: 2007, 'Quasi-static evolution for fatigue debonding'. *ESAIM Control Optim. Calc. Var.* To appear.

Fonseca, I. and N. Fusco: 1997, 'Regularity results for anisotropic image segmentation models'. *Ann. Scuola Norm. Sup. Pisa Cl. Sci.* **4**(3), 463–499.

Francfort, G. A. and A. Garroni: 2006, 'A Variational View of Partial Brittle Damage Evolution'. *Arch. Ration. Mech. An.* **182**(1), 125–152.

Francfort, G. A. and C. Larsen: 2003, 'Existence and convergence for quasi-static evolution in brittle fracture'. *Commun. Pur. Appl. Math.* **56**(10), 1465–1500.

Francfort, G. A., N. Le, and S. Serfaty: 2008, 'Critical Points of Ambrosio-Tortorelli converge to critical points of Mumford-Shah in the one-dimensional Dirichlet case'. To appear.

Francfort, G. A. and J.-J. Marigo: 1998, 'Revisiting brittle fracture as an energy minimization problem'. *J. Mech. Phys. Solids* **46**(8), 1319–1342.

Francfort, G. A. and A. Mielke: 2006, 'Existence results for a class of rate-independent material models with nonconvex elastic energies'. *J. Reine Angew. Math.* **595**, 55–91.

Garrett, K. W. and J. E. Bailey: 1977, 'Multiple transverse fracture in 90° cross-ply laminates of a glass fiber-reinforced polyester'. *J. Mater. Sci.* **12**, 157–168.

Giacomini, A.: 2005a, 'Ambrosio-Tortorelli approximation of quasi-static evolution of brittle fractures'. *Calc. Var. Partial Dif.* **22**(2), 129–172.

Giacomini, A.: 2005b, 'Size effects on quasi-static growth of cracks'. *SIAM J. Math. Anal.* **36**(6), 1887–1928 (electronic).

Giacomini, A. and M. Ponsiglione: 2003, 'A discontinuous finite element approximation of quasi-static growth of brittle fractures'. *Numer. Func. Anal. Opt.* **24**(7-8), 813–850.

Giacomini, A. and M. Ponsiglione: 2006, 'Discontinuous finite element approximation of quasistatic crack growth in nonlinear elasticity'. *Math. Mod. Meth. Appl. S.* **16**(1), 77–118.

Griffith, A.: 1920, 'The phenomena of rupture and flow in solids'. *Philos. T. Roy. Soc. A* **CCXXI-A**, 163–198.

Gurtin, M.: 2000, *Configurational forces as basic concepts of continuum physics*, Vol. 137 of *Applied Mathematical Sciences*. New York: Springer-Verlag.

Hahn, H.: 1914, 'Über Annäherung an Lebesgue'sche Integrale durch Riemann'sche Summen'. *Sitzungsber. Math. Phys. Kl. K. Akad. Wiss. Wien* **123**, 713–743.

Halphen, B. and Q. S. Nguyen: 1975, 'Sur les matériaux standards généralisés'. *J. Mécanique* **14**(1), 39–63.

Hashin, Z.: 1996, 'Finite thermoelastic fracture criterion with application to laminate cracking analysis'. *J. Mech. Phys. Solids* **44**(7), 1129–1145.

Hull, D.: 1981, *An introduction to composite materials*, Cambridge Solid State Science Series. Cambridge University Press.

Irwin, G.: 1958, 'Fracture'. In: *Handbuch der Physik, herausgegeben von S. Flügge. Bd. 6. Elastizität und Plastizität.* Berlin: Springer-Verlag, pp. 551–590.

Jaubert, A.: 2006, 'Approche variationnelle de la fatigue'. Thèse de doctorat, Université Pierre et Marie Curie, Paris.

Jaubert, A. and J.-J. Marigo: 2006, 'Justification of Paris-type fatigue laws from cohesive forces model via a variational approach'. *Continuum Mech. Therm.* **18**(1-2), 23–45.

Kohn, R. and P. Sternberg: 1989, 'Local minimizers and singular perturbations'. *Proc. Roy. Soc. Edin. A* **111**(A), 69–84.

Landau, L. D. and E. M. Lifschitz: 1991, *Lehrbuch der theoretischen Physik ("Landau-Lifschitz"). Band VII.* Berlin: Akademie-Verlag, seventh edition. Elastizitätstheorie. [Elasticity theory], Translated from the Russian by Benjamin Kozik and Wolfgang Göhler, Translation edited by Hans-Georg Schöpf, With a foreword by Schöpf and P. Ziesche.

Lawn, B.: 1993, *Fracture of Brittle Solids - Second Edition*, Cambridge Solid State Science Series. Cambridge: Cambridge University press.

Leblond, J.-B.: 2000, *Mécanique de la rupture fragile et ductile*, Collection Études en mécanique des matériaux et des structures. Paris: Editions Lavoisier.

Leguillon, D.: 1990, 'Calcul du taux de restitution de l'énergie au voisinage d'une singularité'. *C. R. Acad. Sci. II b* **309**, 945–950.

Lemaître, J. and J.-L. Chaboche: 1985, *Mécanique des Matériaux Solides.* Dunod, Paris.

Leonetti, F. and F. Siepe: 2005, 'Maximum principle for vector valued minimizers'. *J. Convex Anal.* **12**(2), 267–278.

Lorentz, E. and S. Andrieux: 1999, 'A variational formulation for nonlocal damage models'. *Internat. J. Plast.* **15**, 119–138.

Marigo, J.-J.: 2005, 'Tearing of a heterogeneous slab'. Unpublished.

Marigo, J.-J. and L. Truskinovsky: 2004, 'Initiation and propagation of fracture in the models of Griffith and Barenblatt'. *Continuum Mech. Therm.* **16**(4), 391–409.

Mielke, A.: 2005, 'Evolution of rate-independent systems'. In: *Evolutionary equations*, Vol. II of *Handb. Differ. Equ.* Elsevier/North-Holland, Amsterdam, pp. 461–559.

Mumford, D. and J. Shah: 1989, 'Optimal approximations by piecewise smooth functions and associated variational problems'. *Commun. Pur. Appl. Math.* **XLII**, 577–685.

Murat, F.: 1985, 'The Neumann sieve'. In: *Nonlinear variational problems (Isola d'Elba, 1983)*, Vol. 127 of *Res. Notes in Math.* Boston, MA: Pitman, pp. 24–32.

Needleman, A.: 1992, 'Micromechanical modelling of interface decohesion'. *Ultramicroscopy* **40**, 203–214.

Negri, M.: 1999, 'The anisotropy introduced by the mesh in the finite element approximation of the Mumford-Shah functional'. *Numer. Func. Anal. Opt.* **20**(9-10), 957–982.

Negri, M.: 2003, 'A finite element approximation of the Griffith model in fracture mechanics.'. *Numer. Math.* **95**, 653–687.

Negri, M. and M. Paolini: 2001, 'Numerical minimization of the Mumford-Shah functional'. *Calcolo* **38**(2), 67–84.

Nguyen, O., E. A. Repetto, M. Ortiz, and R. A. Radovitzki: 2001, 'A cohesive model of fatigue crack growth'. *Int. J. Frac. Mech.* **110**, 351–369.

Nguyen, Q. S.: 2000, *Stability and Nonlinear Solid Mechanics.* London: Wiley & Son.

Paris, P. C., M. P. Gomez, and W. E. Anderson: 1961, 'A rational analytic theory of fatigue'. *The Trend in Engineering* **13**(8), 9–14.

Roe, K. L. and T. Siegmund: 2002, 'An irreversible cohesive zone model for interface fatigue crack growth simulation'. *Int. J. Frac. Mech.* **70**, 209–232.

Rogers, C. A.: 1970, *Hausdorff measures*. London: Cambridge University Press.

Rudin, W.: 1973, *Functional Analysis*. New York: Mc Graw Hill.

Suquet, P.: 1982, 'Plasticité et homogénéisation'. Doctorat d'État, Université Pierre et Marie Curie.

Toader, R. and C. Zanini: 2005, 'An artificial viscosity approach to quasistatic crack growth'. Technical report, S.I.S.S.A. http://cvgmt.sns.it/papers/.

Tvergaard, V.: 1990, 'Effect of fiber debonding in a whisker-reinforced metal'. *Mat. Sci. Eng. A-Struc.* **125**, 203–213.